A Survey of Blur Detection and Sharpness Assessment Methods

Synthesis Lectures on Algorithms and Software in Engineering

Editor
Andreas Spanias, *Arizona State University*

Advances in Waveform-Agile Sensing for Tracking
Sandeep Prasad Sira, Antonia Papandreou-Suppappola, and Darryl Morrell
2008

Despeckle Filtering Algorithms and Software for Ultrasound Imaging
Christos P. Loizou and Constantinos S. Pattichis
2008

A Survey of Blur Detection and Sharpness Assessment Methods

Juan Andrade

ISBN: 978-3-031-00401-8   paperback
ISBN: 978-3-031-01529-8   ebook
ISBN: 978-3-031-00017-1   hardcover

DOI 10.1007/978-3-031-01529-8

A Publication in the Springer series
SYNTHESIS LECTURES ON ALGORITHMS AND SOFTWARE IN ENGINEERING

Lecture #20
Series Editor: Andreas Spanias, *Arizona State University*
Series ISSN
Print 1938-1727   Electronic 1938-1735

A Survey of Blur Detection and Sharpness Assessment Methods

Juan Andrade
University of Cuenca

SYNTHESIS LECTURES ON ALGORITHMS AND SOFTWARE IN ENGINEERING #20

ABSTRACT

Blurring is almost an omnipresent effect on natural images. The main causes of blurring in images include: (a) the existence of objects at different depths within the scene which is known as defocus blur; (b) blurring due to motion either of objects in the scene or the imaging device; and (c) blurring due to atmospheric turbulence.

Automatic estimation of spatially varying sharpness/blurriness has several applications including depth estimation, image quality assessment, information retrieval, image restoration, among others.

There are some cases in which blur is intentionally introduced or enhanced; for example, in artistic photography and cinematography in which blur is intentionally introduced to emphasize a certain image region. Bokeh is a technique that introduces defocus blur with aesthetic purposes. Additionally, in trending applications like augmented and virtual reality usually, blur is introduced in order to provide/enhance depth perception.

Digital images and videos are produced every day in astonishing amounts and the demand for higher quality is constantly rising which creates a need for advanced image quality assessment. Additionally, image quality assessment is important for the performance of image processing algorithms. It has been determined that image noise and artifacts can affect the performance of algorithms such as face detection and recognition, image saliency detection, and video target tracking. Therefore, image quality assessment (IQA) has been a topic of intense research in the fields of image processing and computer vision. Since humans are the end consumers of multimedia signals, subjective quality metrics provide the most reliable results; however, their cost in addition to time requirements makes them unfeasible for practical applications. Thus, objective quality metrics are usually preferred.

KEYWORDS

defocus blur, motion blur, blur detection, out-of-focus, sharpness, blur metrics, image processing, video processing, image quality assessment, image enhancement

To my wife, Emma; my children, Pamela, Karen, and Juan David; my parents, José and Laura; and my friends, Joe and Tracy.

xi

Contents

Preface

The book was inspired by the author's research in the SenSIP Center, the Image, Video, and Usability Lab (IVU Lab), and the School of Electrical, Computer, and Energy Engineering at Arizona State University. The book is intended to be a comprehensive literature review of classic and state-of-the-art methods on blur detection and no-reference blur assessment. Portions of this survey and tutorial book are based on research experiences gained by working with faculty and students at Arizona State University including: Drs. Lina Karam, Pavan Turaga, and Andreas Spanias.

The book is organized in five chapters. Chapter 1 presents a brief introduction to image blurring, different sources of blurring, and the consequences of blurring in image processing algorithms. In Chapter 2, the theory behind several methods devoted to detect out-of-focus blur is presented in detail, and these methods are evaluated using publicly available datasets. In Chapter 3, a survey of full-reference methods for sharpness assessment is presented as well as an introduction to reduced-reference sharpness assessment methods. In Chapter 4, no-reference methods for sharpness assessment are analyzed in detail. Chapter 5 presents a summary and future research directions.

Juan Andrade
December 2020

Acknowledgments

First and foremost, I would like to express my deepest thanks to Professor Andreas Spanias for the opportunity and encouragement in writing this book; his advice and suggestions have helped to improve the content and structure of the present project. I would also like to express my gratitude to Professor Pavan Turaga for his guidance and support during my doctoral studies at Arizona State University.

The author has been supported in part from SenSIP Center at Arizona State University, the Ecuadorian Secretariat of Science, Technology, and Innovation (SENESCYT), and the University of Cuenca.

Juan Andrade
December 2020

CHAPTER 1

Introduction

In this book, we present a survey of the state-of-the-art in image blur detection and assessment. The dimension reduction caused when acquiring an image is accompanied by distortions introduced by the imaging system, objects in the scene, environmental conditions, lighting conditions, among others. Perhaps one of the most prevalent effects produced by these agents is image blurring. Although in most applications image blur is an unwanted effect, there are a few cases where it is intentionally introduced or even enhanced, e.g., in art, photography, and cinematography [1]. The most common sources of blur are out-of-focus, motion, and atmospheric turbulence. These are described below.

1.1 OUT-OF-FOCUS BLUR

Out-of-focus blur, also known as defocus blur, is created by objects at different depths in the scene. The amount of defocus blur in an image depends on the depth-of-field (DoF) of the imaging device. The DoF is defined as the distance between the nearest and furthest objects that present imperceptible blurriness. The DoF can be explained using the thin lens model [2]. Figure 1.1 shows the thin lens model scheme where two objects located at equal distances from the focal plane and over the optical axis can reproduce a circle of confusion with the same radius which translates into a depth ambiguity.

There are several cues that have been used in order to enhance the depth perception in drawings and paintings, e.g., perspective (parallel lines will merge in the infinite) and occlusion (objects closer to the viewer occlude objects that are farther). Also, it has been shown that the degree of blur at the interface between blur and sharp regions is used by the Human Vision System (HVS) as a depth cue [3]. In Fig. 1.2 we reproduced the examples presented in [3]. It can be seen that the inclusion of blurring in the background of images, either (a) natural or (b) synthetic, can create or increase the perception of depth. In both cases, a Gaussian mask with $\sigma = 3$ was used to blur the background.

1.2 MOTION BLUR

Motion blur can be produced either by the motion of objects in the scene which is usually referred to as motion blur or movement of the imaging system, a.k.a. shaking blur. Motion and shaking blur can be reduced by increasing the shutter speed, i.e., by reducing the acquisition time; however, this has the drawback to reduce the Signal to Noise Ratio (SNR) of the image.

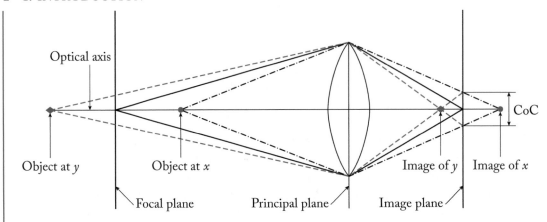

Figure 1.1: Thin lens model scheme used for explaining the out-of-focus image creation.

To compensate for the SNR we can increase the illumination conditions, but there are several applications where the illumination cannot be manipulated such as astronomy, defense, and when imaging photoluminescent elements [4]. It is known that the HVS adds signals over time, a condition that increases the visual sensitivity, but it also introduces motion smearing when looking at moving objects. Motion smearing has been exploited artistically in still images to provide the sensation of motion to the observer [4].

The HVS summates stimulus for over 120 ms in daylight conditions,[1] producing an enhancement of the vision under low-intensity conditions [5]; however, this also can cause the smearing, a.k.a. motion blur, if the object in the scene or the observer is under motion [4]. Although smearing produced by motion is usually considered an artifact that needs to be removed from images, it can also be used to estimate motion [4]. In fact, based on the observation that the human perception of motion smear is high under small observations of motion (up to 20–30 ms) but the perception decreases when the observation is longer; it has conjectured that the HVS uses the information embedded in the motion to estimate the motion parameters and then reduces the amount of blur by motion (deblurring).

Determining the direction of motion can be solved to some extent using the motion streaks left by a moving object. Motion streaks, also known as speed lines, have been widely used for cartoonists to indicate motion [6]. However, motion-sensing ambiguities as those produced by the spatial extent, a.k.a. aperture problem, can mislead the determination of the direction of motion; see an example of the barber-pole illusion in [7].

Based on the fact that moving objects look more blurred in brief ($\sim$ 40 ms) rather than in long exposures ($\sim$ 120 ms), it has been suggested that the HVS has some mechanism for suppressing motion blur.

---

[1]Animals like the toad (Bufo Bufo) summates retinal stimulus for 1–2.5 s, allowing them to see in dim conditions [5].

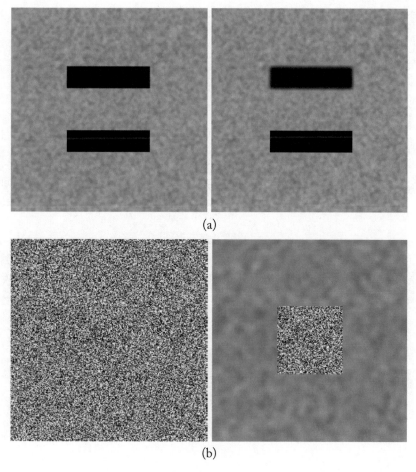

(a)

(b)

Figure 1.2: Example of the effect of blurring in the perception of depth in (a) the left image has one of the bars blurred which is perceived as a different depth and (b) the left image has been blurred except in the central region, although the image shows a random pattern, the perception of depth is notorious. Both images were blurred using a Gaussian mask with $\sigma = 3$.

1.3   ATMOSPHERIC BLURRING

The refractive index of the atmosphere depends on several factors such as pressure, temperature, humidity, among others [8]. Spatial and temporal fluctuations of these factors can be modeled as dynamic random processes and cause changes in the refractive index of the atmosphere [9]. The troposphere, the lowest layer of the atmosphere, is the most dynamic due mainly to the thermal movement of air. Atmospheric turbulence degrades all kinds of optical applications like optical

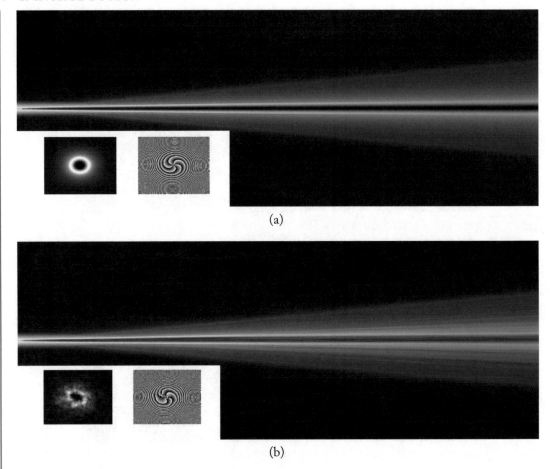

Figure 1.3: Propagation simulation of several rays under atmosphere (a) without turbulence and (b) with turbulence. Images reproduced from [9].

astronomy [10], optical open-air communications [11], and long-distance imaging near to the ground [8, 12].

Imaging a point through a channel with fluctuation of the index of refraction causes changes in position, size, and intensity; these effects are usually referred, in the optical astronomy community, as dancing, pulsating, and scintillation, respectively [13]. A common example of the scintillation caused by atmospheric turbulence is the "twinkle" effect of stars. Figure 1.3 presents the simulation of the propagation of several rays through an atmospheric path with and without turbulence; it can be seen that in the non-turbulence case, rays diverge linearly with time.

1.4   OBJECTIVE SHARPNESS METRIC ASSESSMENT

Digital images and videos are produced every day and the demand for higher quality is constantly rising which creates a need for advanced image quality assessment. Although subjective quality metrics provide the most reliable results, their cost, in addition to time requirements, makes them unfeasible for practical applications. Therefore, objective quality metrics (OQM) are usually preferred. OQM are categorized according to the volume of information required about the ground truth image in full-reference (FR), reduced-reference (RR), and no-reference (NR).

No-reference objective quality methods are preferred since they do not require having access to the pristine image which is unavailable in most practical applications such as image and video coding, texture preservation, image forgery, among others.

1.5   TARGETED APPLICATIONS

Blur detection can be used as a previous step to several image processing and computer vision algorithms such as image deblurring [14], image enhancement [15], refocus [16], depth estimation [17], saliency detection [18], among others.

Additionally, understanding the blurring process can be useful when dealing with applications where the addition of blurring is required in order to provide a realistic feeling like augmented reality and augmented virtual [19].

1.6   ORGANIZATION

The book is organized as follows. Chapter 2 provides a survey of different approaches used to obtain blur features. Since blurring can have different sources such as difference in depth, motion, and atmospheric conditions, it is not possible to create a single blur feature able to describe all kinds of blurring. Therefore, several blur features are usually combined in order to create a blur map of an input image. Most relevant manuscripts about blur map definition are explained, evaluated, and compared. Additionally, a new method proposed by the author is described in detail. Chapter 3 presents a review of relevant image quality assessment (IQA) techniques as a function of the amount of information available about the pristine image. A survey of full-reference sharpness assessment methods is presented as well as an introduction to reduced-reference sharpness assessment methods. In Chapter 4, special attention is devoted to no-reference image quality assessment due to the number of possible applications. Chapter 5 presents a summary of the book. At the end we provide an extensive bibliography.

CHAPTER 2

Out-of-Focus Blur

Natural images usually include defocus blur due to the existence of objects at different depths from the camera. The depth richness of a scene translates into a spatially variable defocus blur in the captured image which cannot be easily undone with image deconvolution algorithms not only due to their computational requirements but also because most of the blind deconvolution algorithms assume a spatially invariant blur [20]. Automatic blur detection is an important element for several computer vision tasks such as spatially varying deblurring [21], photo editing [22], image classification [23], depth estimation [24], saliency detection [18], image segmentation [25], and digital image forensic analysis [26].

In order to measure the perceived image quality, which impacts the user's Quality of Experience (QoE), blurriness metrics were first proposed [27]. The consolidation of the blur information from the whole image into a single score has several applications like image denoising [29], image compression [30], image super-resolution [31], among others. Single score index methods are employed in situations where the blurring is considered spatially homogeneous. For the spatially varying blurring, on the other hand, blur must be determined using local information, i.e., much less information.

Blur detection methods can be divided into two categories: those using several images [32, 33] and those using a single image [16, 17, 25, 34–37, 39–41, 43, 44, 114]. The use of several images is restrictive since multiple images of the same scene but under different constraints, e.g., different shutter speeds are required; that is why single image blur detection has been a strong interest of the research community. Blurred regions in an image are usually referred to as background because they usually do not represent interest for the viewer; on the other hand, sharp regions are usually referred to as foreground. For example, in Fig. 2.1 the boy and the butterfly are the foreground, meaning the regions of interest of the images and the background, which occupies the larger region looks blurred.

The identification of sharp regions can be useful to apply more sophisticated and image understanding algorithms [46] such as feature detection [47], generation of image descriptors [48], scene understanding [49], among others. Similarly, deconvolution algorithms [50–52] can be applied to detected blur regions.

According to the approach used, blur detection algorithms can be classified as gradient-based [36, 37], intensity-based [53], perceptual-based [16, 54], and transform-based [17, 34, 40]. Additionally, to improve blur detection, most of the state-of-the-art algorithms in blur detection are multi-feature and multi-scale [36, 37, 40].

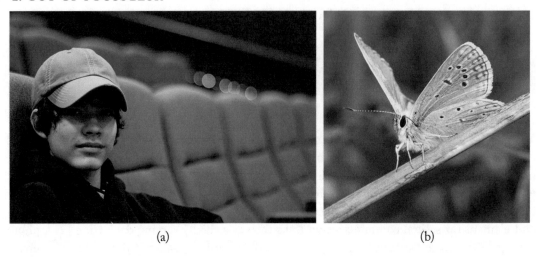

(a)   (b)

Figure 2.1: Example of defocus blur created due to objects at different depths in the scene; images taken from [38]. Note how image blurring increases with the distance from the focal plane, i.e., the sharp part of the image.

The core of every blur detection algorithm are the blur features created using local information. Blur features need to be descriptive enough to distinguish small levels of blurriness under different levels of luminance, color, scale, and texture. Therefore, usually, several blur features are combined to create a blur map.

2.1   GRADIENT-BASED BLUR DETECTION

It is widely accepted that image blur reduces high-frequency energy; therefore, when reblurring an image that has in it sharp and blurred regions will pop-out sharp regions since those areas will show a bigger change with respect to the initial image, i.e., blur regions are more invariant to a low-pass filtering process [55]. In Fig. 2.2, we show the result of reblurring for the detection of sharp regions. The drawback of this light and straightforward approach is that textureless regions will be portrayed as blurred regions; see the cheeks of the lady in the picture.

In the image restoration community, the gradient of images has been studied and successfully used as a regularization prior. Assuming a Gaussian distribution of image gradients provides a closed-form solution but a smooth restored image, i.e., high-frequency features such as edges are not well reconstructed [56]. Later, Laplacian distribution was used as the probability density function (PDF) of image gradients. This improved the quality of the reconstructed image substantially, however, iterative algorithms were necessary to solve the optimization problem [52, 57]. Recent studies have shown that gradients of natural images are better described using a hyper-Laplacian distribution, i.e., $P(x) \propto e^{-\lambda |x|^{\alpha}}$ where $0.5 \leq \alpha \leq 0.8$ [58]. Figure 2.3

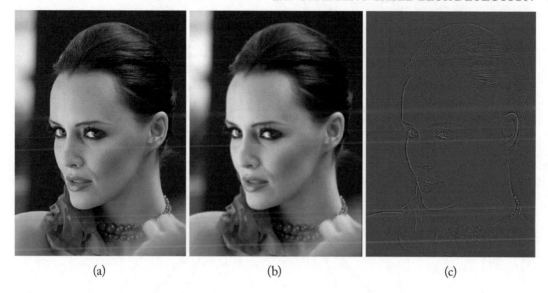

(a)   (b)   (c)

Figure 2.2: Use of reblurring for detecting sharp regions: (a) input image, (b) reblurred image using a Gaussian kernel with $\sigma = 1$, and (c) subtraction of the input and reblurred images, clearly the sharp regions in the input image are better preserved, however textureless regions such as cheeks are reported as blurred regions. Image taken from the database [38].

shows the average gradients' PDF computed using the 200 testing images of the BSDS500 dataset [59], and the PDFs defined using Gaussian, Laplacian, and Hyper-Laplacian priors; clearly, the hyper-Laplacian prior provides a representation tighter to the real PDF.

2.1.1 METHODS USING KURTOSIS

Based on the wrong belief that kurtosis is a measure of the peakedness of a frequency distribution [60], it has been widely used as a sharpness measure. Its good performance is linked to the fact that kurtosis measures the tailedness of a PDF, and blurred images present PDFs with smaller tails due to the attenuation/elimination of higher-frequency content; see Fig. 2.4.

In [37, 61] kurtosis is defined as

$$K(a) = \frac{E[a^4]}{E[a^2]} - 3,$$

(2.1)

where $E[\cdot]$ is the expectation operator and a is the input data vector, used as sharpness metric.

In [61], kurtosis is used to create a global sharpness index. The average of the kurtosis of the discrete cosine transform (DCT) of 8×8 pixels windows centered in pixels flagged as being part of edges [62] is used as a global image sharpness index.

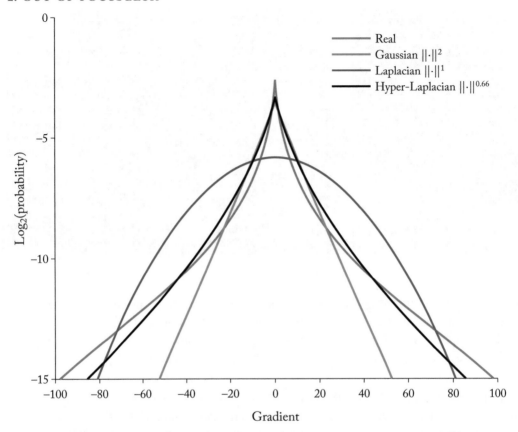

Figure 2.3: Comparison of PDFs used as gradient image priors. The average gradients' probability density function (PDF) computed using the 200 testing images of the BSDS500 dataset [59], and the PDFs defined using Gaussian, Laplacian, and Hyper-Laplacian priors are shown. Clearly, the hyper-Laplacian prior provides a representation tighter to the real one.

2.2   MULTI-FEATURE METHODS

In order to reach better performance, state-of-the-art algorithms use multiple features, e.g., frequency-based, gradient-based, luminance-based, among others. The features used for prominent methods, as well as the fusion method used, is presented and analyzed in this section.

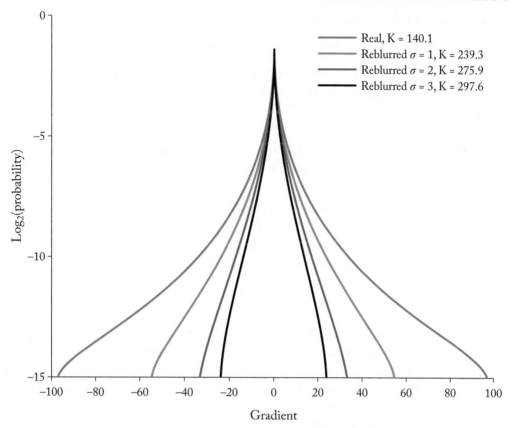

Figure 2.4: Loss of high-frequency content in natural images under reblurring. The average gradients' PDF computed using the 200 testing images of the BSDS500 dataset [59], and the PDFs defined using reblurred versions of the input images are presented. The kurtosis can detect bluriness because the tailedness of the distributions change as a function of blur and not due to the peakedness.

2.2.1   KURTOSIS-, GRADIENTS' DISTRIBUTION-, AND SPECTRAL SLOPE-BASED METHODS

In [37] a multi-feature is proposed to create an image sharpness map. Kurtosis is used as one of the sharpness features as follows:

$$f_1 = \min\left(\ln K(B_x) + 3, \ln K(B_y) + 3\right), \tag{2.2}$$

where B_x and B_y are the horizontal and vertical gradients of the input image, respectively. The second sharpness feature used in [37] considers the tailedness of the gradients' distributions of the input image. A Gaussian mixture model (GMM) [72] with two components is fit to the

PDF of the image gradients as follows:

$$\Delta B \sim \pi_1 \mathcal{N}(\mu_1, \sigma_1) + \pi_2 \mathcal{N}(\mu_2, \sigma_2). \tag{2.3}$$

Considering $\sigma_1 > \sigma_2$, i.e., σ_1 is associated with the Gaussian that describes the tails of the original PDF, the second sharpness feature is defined as

$$f_2 = \sigma_1. \tag{2.4}$$

The third sharpness feature comes from the frequency domain. It is known that the average power spectrum of natural images of Eq. (2.5) decays with the square of the frequency, i.e., $1/\omega^2$ [73]:

$$J(\omega) = \frac{1}{n} \sum_{\theta} |\mathcal{F}(\omega, \theta)|^2, \tag{2.5}$$

where n is the number of angular directions θ and $\mathcal{F}$ represents the Fourier transform (FT). Finally, the sharpness feature based in frequency content is given by

$$f_3 = \sum_{\omega} \log (J(\omega)). \tag{2.6}$$

Besides the features based on image statistics, a set of linearly independent filters is learned. It is shown that the learned filters have a high-pass characteristic. The posterior score of a naive Bayesian classifier [74] that combine all the aforementioned features is used as the final sharpness score per scale. In order to combine intra- and inter-scale information [75, 76], loopy belief propagation [77] is used to solve a cost function that includes a data-fitting term, an intra-scale, and an inter-scale term.

2.2.2   GRADIENTS-, SPECTRA-, AND INTENSITY-BASED METHODS

In [36] features representing the gradients, spectra, and color information are combined to detect blurred regions.

The feature that involves gradient information has the same structure as the one used in [37], i.e., a GMM of two components is used to fit the distribution of the gradients; see Eq. (2.3) and the bigger variance out of the two components is used to define the local sharpness metric as follows:

$$f_g = \frac{25\sigma_1}{C_p + \epsilon}, \tag{2.7}$$

where ϵ is a small number to avoid division by zero, and C_p is the normalized contrast of the image block which is given by

$$C_p = \frac{L_{max} - L_{min}}{L_{max} + L_{min}}, \tag{2.8}$$

where L_{max} and L_{min} are the maximum and minimum contrast within the image block under analysis, respectively.

The spectra-based feature comes from the average power spectrum as defined in [37]; see Eq. (2.5). The power α of the expression $J(\omega) \approx A/f^\alpha$ is defined for the whole image α_o and for each image block α_p, the local sharpness metric based on the average power spectrum is given by

$$f_s = \frac{\alpha_p - \alpha_o}{\alpha_o}. \tag{2.9}$$

The feature that involves color saturation is given by

$$f_{sc} = \frac{\max(S_p) - \max(S_o)}{\max(S_o)}, \tag{2.10}$$

where S_p and S_o are the saturation defined for the image block and the whole image, respectively. The expression for S_p is given as follows:

$$f_{cs} = 1 - \frac{3}{R + G + B}[\min(R, G, B)], \tag{2.11}$$

where R, G, and B are the image color planes.

2.2.3   LOCAL AUTOCORRELATION CONGRUENCY-BASED METHOD

In [36] the blur is classified as out-of-focus and motion blur by using local autocorrelation congruency (LAC). Basically, the color spread of pixels to their neighboring pixels is estimated using a weighted convolution which resembles the theory behind the Harris corner detector [78]. The eigenvalues $\lambda_1, \lambda_2, \lambda_1 \geq \lambda_2$ of the matrix M computed as follows

$$M = \sum_{(x_k, y_k) \in W} \begin{bmatrix} I_x^2(x_k, y_k) & I_x(x_k, y_k)I_y(x_k, y_k) \\ I_x(x_k, y_k)I_y(x_k, y_k) & I_y^2(x_k, y_k) \end{bmatrix} \tag{2.12}$$

are used to compute the quotient $\sqrt{(\lambda_1/\lambda_2)}$ which is defined for several orientations θ; as shown in Fig. 2.5. The so-created histogram has a stronger response when the blur is due to motion which corresponds to the response to an edge in terms of Harris corner detector. Finally, for both, blur detection and blur classification Bayes classifiers are trained.

2.2.4   SPECTRAL- AND SPATIAL-BASED METHODS

In [41] a sharpness map is defined using spectral and spatial properties of the input image. The created sharpness map can be collapsed into a single value which can be used as a global sharpness metric of the image. The input color image is first converted to grayscale. Then, based on the observation of [79] which states: "*an image region whose magnitude spectrum exhibits a slope factor of $0 \leq \alpha \leq 1$ will appear sharp, whereas regions with $\alpha > 1$ will appear progressively less sharp as α increases,*" the α exponent of the magnitude spectrum is evaluated for each image block as follows.

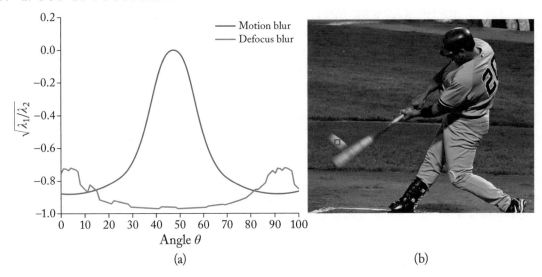

Figure 2.5: Example of blur classification. (a) Plot of $\sqrt{(\lambda_1/\lambda_2)}$ for different gradient orientations. Two blocks are analyzed one under motion blur and the other affected by defocus blur; the input image taken from the database [38] is shown in (b).

An image block y of 32×32 pixels is declared as having zero contrast if one of the following conditions is fulfilled:

$$\overline{(b + ky)^\gamma} \leq 2$$
$$\max\left((b + ky)^\gamma - \min(b + ky)^\gamma\right) \leq 5, \tag{2.13}$$

where $\overline{[\cdot]}$ represents the mean, and $b = 0.7656$, $a = 0.0364$. Only image blocks that do not fall within these conditions are processed. The parameter α is estimated by fitting a line of the form $-\alpha \log(f) + \log(\beta)$ to the magnitude spectrum summed across all orientations as follows:

$$\alpha_y = \underset{\alpha}{\operatorname{argmin}} \left\| \beta f^{-\alpha} - \sum_\theta |\mathcal{F}\{y\}(f, \theta)| \right\|_2^2. \tag{2.14}$$

Finally, the sharpness component based on the magnitude spectrum of block y is defined applying a sigmoid function as follows:

$$S_1(y) = 1 - \frac{1}{1 + e^{-3(\alpha_y - 2)}}. \tag{2.15}$$

The second component considers the perceived sharpness. For each full-overlapping image block y_2 of 8×8 pixels, the perceived sharpness is estimated as

$$S_2(y_2) = \frac{1}{4} \max_{\zeta \in y_2} v(\zeta), \tag{2.16}$$

Figure 2.6: Example of performance of the algorithm proposed by [41]. (a) Input image taken from the database [42], (b), and (c) present the results of S_1 and S_2, respectively; (d) shows the final sharpness map as expressed by Eq. (2.17).

where ζ are 2×2 pixel blocks, and $v(\zeta)$ is the total variation of ζ. Both sharpness features S_1 and S_2 are combined using a weighted geometric mean as follows:

$$S_3 = S_1^{0.5} \times S_2^{0.5}.$$

(2.17)

Figure 2.6 shows both sharpness features proposed in [41] as well as the result.

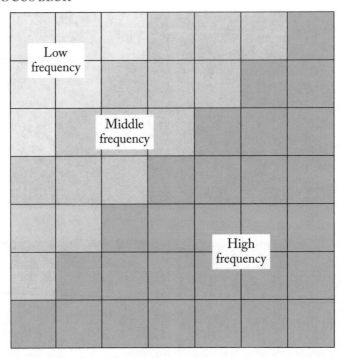

Figure 2.7: Segmentation of DCT coefficients used in [17], where only the high-frequency coefficients are used.

2.3 TRANSFORM-BASED METHODS

2.3.1 THE DISCRETE COSINE TRANSFORM (DCT)-BASED METHOD

In [17], a spatially varying blur detection method based on the high-frequency (HF) multiscale fusion and sort transform (HiFST) is proposed. The algorithm uses full overlapping blocks of $M \times M$ pixels of the gradients' magnitude of the input image. The DCT is computed for every block and the HF coefficients defined, as shown in Fig. 2.7, are collected for every pixel. In order to provide the algorithm with a multi-scale characteristic, the described process is repeated for four different block sizes. The multi-scale HF DCT coefficient for each pixel is collected and sorted in ascending order. As a result, $N_{coef} = \sum_{r=1}^{4} \frac{M_r(1+M_r)}{2}$ HF coefficients are available for each image pixel, where M_r are the block sizes used in the multi-scale. Each plane of HF coefficients $L_t, t \in [1, \ldots, N_{coef}]$ are normalized as follows:

$$\hat{L}_t = \frac{L_t - \min L_t}{\max L_t - \min L_t}. \tag{2.18}$$

The blur detection map is defined as $D = T \cdot \omega$, where T is the maximum of all the HF coefficients for each image pixel position, and ω is defined as the local entropy map of T us-

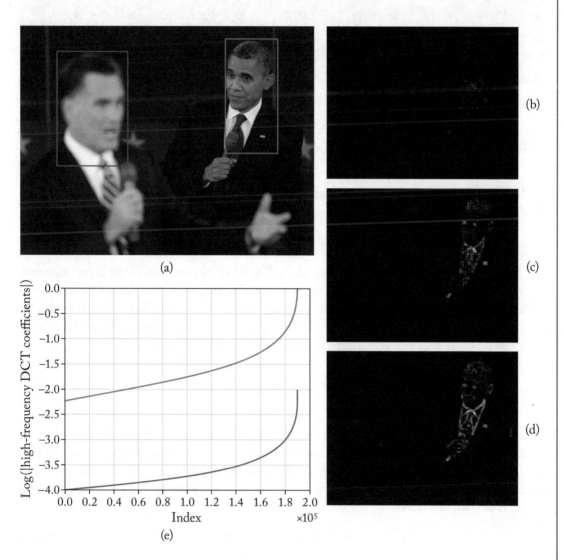

Figure 2.8: Example of performance of the DCT coefficients for detecting blur/sharp regions. (a) Shows the input image taken from the database [38] where two gentlemen, one in focus and one out-of-focus, are portraited. (b)–(d) Show the feature planes formed with the smallest, medium, and highest HF DCT coefficients, respectively. (e) Presents the sorted HF DCT coefficients for the regions framed in (a); clearly, the two regions, blurred and sharp, can easily be segmented.

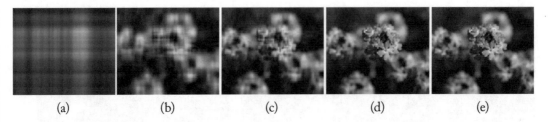

(a)   (b)   (c)   (d)   (e)

Figure 2.9: Example of reconstruction using (a) 1%, (b) 5%, (c) 10%, and (d) 50% of eigenimages; (e) shows the original image taken from the database [38].

ing blocks of 7×7 pixels. Finally, in order to eliminate outliers edge-preserving filtering is applied [81].

2.3.2   SVD-BASED METHODS

In [39] a blur detection and classification method based on the analysis of the image eigenvalues is proposed. Given an image Y it can be expressed using Singular Value Decomposition (SVD) as $Y = U \Lambda V^T$ where U, V are orthogonal matrices and Λ is a diagonal matrix with the eigenvectors arranged in decreasing order. This representation has been widely used for image compression purposes since the image can be reconstructed as follows:

$$Y = \sum_{i=1}^{n} \lambda_i u_i v_i^T,$$ (2.19)

where u_i, v_i, and λ_i are columns of u, v, and the i eigenvalue, respectively. The lower indexes of i, i.e., the bigger eigenvalues correspond to the coarse content of the image (low-frequency content), while small eigenvalues provide fine details associated with high-frequency content. Since blurring attenuates or eliminates high-frequency content of the image, the proposed sharpness metric is defined as follows:

$$S = \frac{\sum_{i=1}^{k} \lambda_i}{\sum_{i=1}^{n} \lambda_i} \quad k < n.$$ (2.20)

The sharpness metric can be understood as a quotient of the low- to high-frequency content of the image. An image pixel is declared as being part of the blurred region if S defined in Eq. (2.20) is higher than a threshold. Blurring on images is classified as motion or defocus blur adopting the method of [35].

Figure 2.9 shows the reconstruction of an image taken from [38] with different a number of eigencomponents, when the addition of eigencomponents associated with smaller eigenvalues increase the level of detail in the reconstructed image, i.e., smaller eigenvalues are associated with high-frequency content of the image.

2.3.3   METHOD BASED ON THE RANK OF LOCAL PATCHES

In [43] based on simulations, it has been observed that the matrix rank of a patch decreases when the patch is blurred using a Gaussian kernel. Four patches with different sampling orientations are used to look for the one that is closest to the diagonal or anti-diagonal, i.e., full rank. Only edge pixels are used to define an initial sparse sharpness map and later image matting Laplacian [82] is used to propagate the sharpness information to regions sharing the same color information. The algorithm of the method [43] is presented in Algorithm 2.1.

Algorithm 2.1 Blur map detection based on rank of local patches.

1: Given Single image affected by defocus blur I produces a Defocus map S.
2: Set d
3: Compute the gradient image $G = \nabla I$ where: $\nabla = \frac{\partial}{\partial x} + \frac{\partial}{\partial y}$
4: Define edge positions I_e on the input image using Canny edge detector [62]
5: **for all** Edge pixels detected $I_e(i_e, j_e)$ **do**
6:     Sample four gradient patches as follows:
7:     $Q_1(i, j) = G(i_e - p + i, j_e - p + j)$
8:     $Q_2(i, j) = G(i_e - p + j, j_e + p - i)$
9:     $Q_3(i, j) = G(i_e + i - j, j_e - p + i)$
10:     $Q_4(i, j) = G(i_e - p + i, j_e - i + j)$
11:     Compute the matrices $P_k = Q_k + Q_k^T$, $k \in [1, \ldots, 4]$
12:     The blur map value at the edge pixel position is given by:
13:     $S(i_e, j_e) = -\ln(1 - \frac{1}{2p+1} \max P_k)$ $k \in [1, \ldots, 4]$
14: **end for**
15: Apply image matting Laplacian [82] to the generated sparse blur map. Matting Laplacian propagates the blur information defined for edge positions to other image regions using color information.

Sample results of the sparse defocus map defined using the rank of patches of the image gradients around edge pixels are shown in Fig. 2.10. The results provided by the matting Laplacian [82] algorithm is highly dependent on the initialization.

2.3.4   LOCAL BINARY PATTERNS-BASED METHOD

In [25], local binary patterns (LBP) is proposed as a sharpness metric. LBP was first proposed in [83], and has been successfully used in several image processing and computer vision applications [84, 85]. Given the number of neighbors P, the radius of the circle R, and considering the central pixel $g_c(x_c, y_c)$ located at $(0, 0)$, the position of the neighbor pixels are given by $(-R \sin(\frac{2\pi p}{P}), R \cos(\frac{2\pi p}{P}))$. Figure 2.11 presents three examples of the circularly symmetric constellation sets for different values of (P, r).

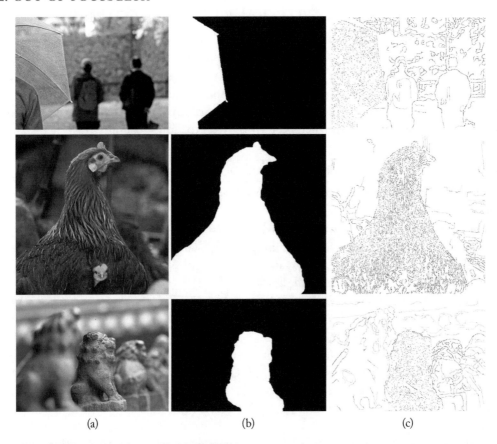

(a)   (b)   (c)

Figure 2.10: Sample results from [43]. (a)–(b) Images and ground-truth taken from the database [38]. (c) Sparse defocus map defined using the rank of gradient patches.

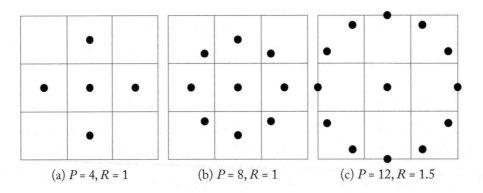

(a) $P = 4, R = 1$ (b) $P = 8, R = 1$ (c) $P = 12, R = 1.5$

Figure 2.11: Examples of the circularly symmetric constellation sets for different values of (P, r).

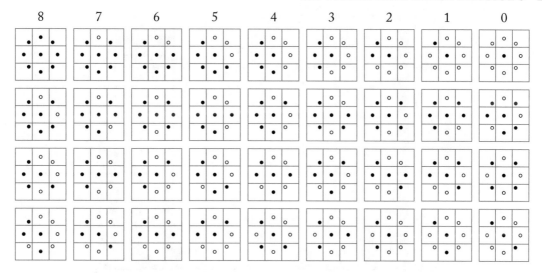

Figure 2.12: Possible patterns for $P = 8$, $R = 1$. Only ten categories exist when considering the definition of uniform patterns.

The PBP code for a central pixel $g_c(x_c, y_c)$ and a pair (P, R) is given by

$$LBP_{P,R}(x_c, y_c) = \sum_{p=0}^{P-1} S(n_p - n_c) \times 2^p, \qquad (2.21)$$

where n_c and n_p are the intensity values of the central and neighboring pixels, respectively. The function $S(x)$ is given as follows:

$$S(x) \begin{cases} 1 & |x| \geq T_{LBP} \\ 0 & |x| < T_{LBP}, \end{cases} \qquad (2.22)$$

where T_{LBP} is a positive threshold empirically determined.

Considering the rotation invariant version of the LBP [85] for $P = 8$, $R = 1$, we have 36 patterns as shown in Fig. 2.12. Considering these patterns as uniform if they do not have more than two changes 1 to 0 or 0 to 1, we can reduce the number of possible patterns to 10. Patterns marked from 0–8 in Fig. 2.12 are uniform patterns and all the remaining patterns form the tenth category.

The frequency distribution of LBP is presented in Fig. 2.13. Bins 0–8 correspond to uniform patterns presented in the first row of Fig. 2.12, while the tenth bin in the distribution corresponds to all the remaining patterns in Fig. 2.12. It is noted that sharp regions tend to have a higher frequency in the upper bins and this consistent since higher bins are associated with a higher number of changes and, thus, higher-frequency content.

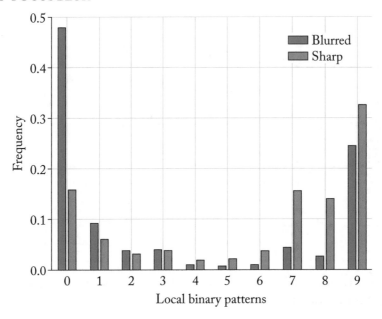

Figure 2.13: Distribution of the LBP codes in blurred and sharp regions of images from [38]. Bins 0–8 are considered uniform patterns and they correspond to the first row of patterns in Fig. 2.12; the tenth bin corresponds to all the remaining patterns in Fig. 2.12.

Based on the frequency distribution of Fig. 2.13, in [25] it is proposed as sharpness metric the normalized sum of LBP with six or more changes as follows:

$$m_{LBP} = \frac{1}{N} \sum_{i=6}^{9} (LBP_{(8,1)}(i)),$$
(2.23)

where $(LBP_{(8,1)}(i))$ is the number of rotation invariant LBP patterns with type i within a ($P = 8, R = 1$) configuration. Figure 2.14 shows the influence of T_{LBP} in the definition of the sharp regions.

2.4   SPARSE REPRESENTATIONS

2.4.1   INTRODUCTION TO SPARSE REPRESENTATION

Sparse representation using learned overcomplete dictionaries have been successfully applied in several image processing applications [86–88]. Although the theory behind sparse representation is not new it has not had much development due mainly to the computational requirements. Sparse representation has a strong link with compressed sensing theory that is why some theorems and definitions are grabbed from this field. Given a linear system $Ax = b$ where

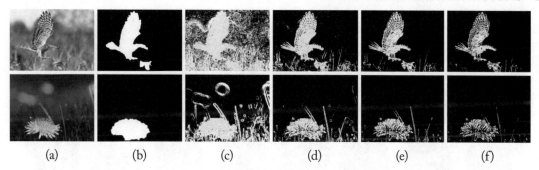

(a) (b) (c) (d) (e) (f)

Figure 2.14: Influence of the threshold T_{LBP} in Eq. (2.22) on the definition of sharp regions. (a)–(b) Input image and the corresponding ground-truth sharpness map taken from the database [38]. (c)–(f) LBP-based sharpness map given by Eq. (2.23) for different values of T_{LBP}: 0.1, 0.3, 0.5, and 0.7, respectively.

$A \in \mathbb{R}^{(n \times m)}$ if $n < m$ then the system is undetermined and it has an infinite number of solutions. A cost function is usually used to narrow down the number of solutions as follows:

$$(P_G): \quad \min_x G(x) \quad \text{subject to} \quad b = Ax. \tag{2.24}$$

The most frequently used cost function is the convex squared ℓ_2 norm, i.e., $G(x) = \|x\|_2^2$ which lead us to the closed-form solution known as pseudo-inverse:

$$\hat{x}_2 = A^+ b = A^T (AA^T)^{-1} b. \tag{2.25}$$

We can also look for a sparse solution, i.e., the solution with only a few nonzero elements. The norm used in this case is the ℓ_0 where $\|x\|_0 = i : x_i \neq 0$; then the ℓ_0 norm penalize equally to all nonzero elements of x. The optimization problem using an ℓ_0 constraint can be expressed as follows:

$$(P_0): \quad \min_x \|x\|_0 \quad \text{subject to} \quad b = Ax. \tag{2.26}$$

One approach to solving the previous equation can be an exhaustive search along the combinatorial problem $\binom{m}{s}$ where m is the number of atoms in A and s represents the sparseness of the solution (number of nonzero elements of the solution); this problem has been shown to be an NP-hard problem.

It has been shown [89] that relaxing the ℓ_0-norm problem by an ℓ_1-norm problem provides a convex functional, that is why a great effort has been devoted to solving the ℓ_1 problem which under certain conditions can provide the solution of the ℓ_0 problem [90]. From the signal processing point of view sparse representations are efficient and widely used, for example given an image I we can expand it over an orthonormal basis Ψ, e.g., FT, DCT, wavelets, etc., where

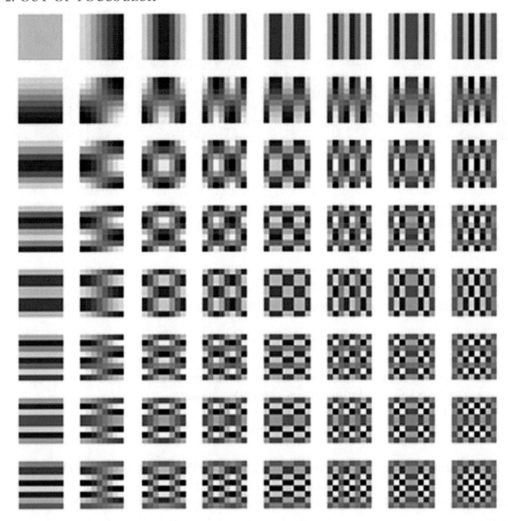

Figure 2.15: 8 × 8 array of discrete cosine transform basis functions.

$\Psi = \psi_1, \psi_2, \ldots, \psi_N$. Using the DCT over an 8 × 8 block can be understood as expanding the vectorized form of the 8 × 8 block, i.e.: f, $f \in \mathbb{R}^n$ (in the example $n = 64$). Considering the basis shown in Fig. 2.15 where every tile corresponds to a basis function in $\mathbb{R}^{64}$. In matrix form the DCT can be written as: $x = f \cdot \Psi$ where $x_i = f \cdot \psi_i$ and each tile of Fig. 2.15 corresponds to a column in Ψ, usually referred as an atom of Ψ.

Now by definition, the recovery is given by $f = \Psi \cdot x$ but if we use x_s instead of x, where x_s is the vector with the same size as x and its coefficients are zero except those that correspond to the highest S values, then an error is included in the reconstruction of the signal, error that even

Figure 2.16: Example of reconstruction using (a) 1%, (b) 5%, (c) 10%, and (d) 50% of DCT coefficients; (e) shows the original Lena image.

under high sparse conditions may be imperceptible or at least acceptable, as shown in Fig. 2.16 where different levels of sparsity have been applied in the reconstruction process of the image. As we can see the ability of DCT or any other transformation, e.g., Discrete Wavelet Transform (DWT) to sparsify the image content is directly related to its compression performance.

In order to check if our solution to the system $A \cdot x \doteq b$ is the sparsest we can use either the *Spark*[1] of A or the concept of mutual coherence [91].

Since the exhaustive search is not a feasible way to find the ℓ_0-norm for $A \cdot x = b$, there are two main streams to solve it, namely matching pursuit and basis pursuit.

2.4.2   GREEDY ALGORITHMS

If we want to find the sparsest solution to P_0 in Eq. (2.26) then we need to seek in the $\binom{N}{K}$ permutations which one fulfills the error requirement. Since we look for the sparsest solution, we have to start with $K = 1$, i.e., $\binom{N}{1}$ if the solution is not 1-sparse, then we have to move forward to search in $\binom{N}{2}$ and so on until we find the one that satisfies the error constraint. However, this is too expensive in computational terms, therefore, greedy approaches have been successfully implemented to solve P_0. Greedy Algorithms (GAs), also known as Matching Pursuit Algorithms (MPs), change the idea of an exhaustive search for the optimal solution by an iterative search of locally optimal terms. GAs have shown to be fast but unfortunately not always able to find the optimum solution.

GAs are not new and have been used in many areas, usually with different names. For example, in statistical modeling greedy stepwise Least-Squares is known as Forward Stepwise Regression [92]; in signal processing, it is called Matching Pursuit (MP) [93, 94] or Orthogonal Matching Pursuit (OMP) [95, 96], approximation theorists call these algorithms as Greedy Algorithms (GAs).

---

[1]The spark of a matrix A [92] is the smallest number of columns from A that are linearly dependent. It is also known as *Kruskal rank* in the psychometrics literature. Given $A \in \mathbb{R}^{(m \times n)}$ we have that $1 \leq spark(A) \leq n + 1$, where $spark(A) = n + 1$ means that no n-columns are linearly dependent.

The main idea behind MP is to take one by one the elements of $\hat{x}$ in such a way that the selected element minimizes the error given by $\|Ax - y\|_2^2$ in other words, the one that maximizes the internal product or the one that is more parallel to the vector b, if the error is below a predefined minimum error ϵ, then that will be the solution otherwise we need an extra iteration in which again we look for the term that minimizes the residual error on b when the selected first term has been subtracted; the process is iterated until some stopping criteria is reached, e.g., the residual error goes below the limit given by ϵ or a maximum number of elements has been reached (K-sparse condition).

MP replaces the exhaustive search of the optimum solution for a series of locally optimal single-terms update. In MP for selecting each element of $\hat{x}$ we have to solve N Least-Squares problems and since the solution is K-sparse we will need to solve a maximum of NK Least-Squares problems.

There are a considerable number of variations of MP among which we have: Stage-wise Orthogonal Matching Pursuit (StOMP) [97], Regularized Orthogonal Matching Pursuit (ROMP) [98], and Compressive Sampling Matching Pursuit (CoSaMP) [99]. In every iteration of OMP the column vector of A that has the highest internal product with the residual vector r is selected as the locally optimum solution.

On the other hand, in StOMP all columns of A whose internal product is above a predefined threshold are selected and used to estimate a possible and the correspondent residual error if the residual error is above the desired error a new iteration is executed. The advantage of StOMP is that it has a faster convergence compared with OMP; but, its results tend to be different as a function of the threshold used to select the vectors from A.

ROMP as StOMP takes in every iteration more than one vector of A to estimate the possible solution for x; but, instead of using a threshold as in StOMP to define the support vectors, ROMP computes the threshold in every iteration as half of the highest internal product, the algorithm is an iterative one as well.

CoSaMP like ROMP and StOMP pick more than one vector from A in each iteration of the algorithm, the difference in CoSaMP is that the number of vectors used to estimate the solution in every iteration is fixed avoiding in this way the threshold of StOMP or the threshold computation used in ROMP.

2.4.3 CONVEX RELAXATION TECHNIQUES

According to [100] we can replace the ℓ_0 norm of Eq. (2.26) by ℓ_1 which has a polynomial time (linear programming), this change is usually known as the "convexation" or "relaxation" of the ℓ_0-norm and is presented in Eqs. (2.27) and (2.28) for the noiseless and noisy cases, respectively:

$$\hat{x} = \underset{b=Ax}{\arg\min} \|x\|_1 \tag{2.27}$$

$$\hat{x} = \underset{\|Ax-y\|_2^2 < \epsilon}{\arg\min} \|x\|_1. \tag{2.28}$$

Relaxation methods, also known as Basis Pursuit (BP), have performance that has been proved theoretically and experimentally when the Restricted Isometry Property (RIP) is satisfied [101]. Although the ℓ_0-norm penalizes the number of nonzero elements on the solution x and the ℓ_1-norm penalizes the sum of the elements of x it has been proved that under certain conditions [102] the ℓ_1 solution is the same as the ℓ_0 approach. The change of norm turns the problem into a convex problem which can be solved in a reasonable time using quadratic programming.

Equations (2.27) and (2.28) can be rewritten as follows:

$$\hat{x} = \underset{\|x\|}{\operatorname{argmin}} \|x\|_1 \quad \text{s.t.} \quad b = Ax \tag{2.29}$$

$$\hat{x} = \underset{\|x\|}{\operatorname{argmin}} \|x\|_1 \quad \text{s.t.} \quad \|Ax - y\|_2^2 < \epsilon. \tag{2.30}$$

There is an important optimization principle known as The Lasso (Least Absolute Shrinkage and Selection Operator) [103] which instead of using the BP method uses a constraint on its value as follows:

$$\hat{x} = \underset{x}{\operatorname{argmin}} \|Ax - b\|_2^2 \quad \text{s.t.} \quad \|x\|_1 \leq \lambda. \tag{2.31}$$

There are many algorithms devoted to solving the aforementioned minimization problems [104] as well as free MATLAB tools such as CVX [105] and ℓ_1-Magic [106].

2.4.4   SPARSE REPRESENTATION-BASED METHOD

In [16] a Just Noticeable Defocus Blur (JNB) detector is proposed. The authors of [16] claim that traditional frequency- and gradient-based methods can perform appropriately when the amount of blur is significant, but they fail when dealing with a small amount of blurring. For the input image Y we take full overlapping blocks of $\sqrt{d} \times \sqrt{d}$ pixels to create the matrix $Y = y_1, \ldots, y_n \in \mathbb{R}^{d \times n}$. Each vectorized block y_i can be expressed using sparse representation as follows:

$$\min_{x_i} \|y_i - Dx_i\|_2^2 \quad \text{s.t.} \quad \|x_i\|_0 \leq k, \tag{2.32}$$

where $D \in \mathbb{R}^{d \times m}$ is an offline trained overcomplete dictionary [107, 108], x_i is the reconstruction vector for y_i and k is a maximum number of non-zero elements in x_i.
Using relation Eq. (2.32) can be written as follows:

$$\min_{x_i} \|x_i\|_1 \quad \text{s.t.} \quad \|y_i - Dx_i\|_2 \leq \epsilon. \tag{2.33}$$

A dictionary of 128 atoms, i.e., $m = 128$ is trained offline using blurred image patched. The sharpness metric for the patch y_i is defined as $f_a = \|x_i\|_0$. Clearly, the sharper the image patch the higher the number of elements required to represent it. Unfortunately, this method

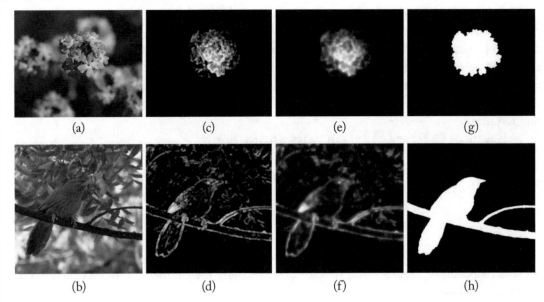

Figure 2.17: Example of performance of the algorithm proposed by [16]. (a)–(b) Input images taken from the database [38], (c)–(d) sharpness map before image matting, (e)–(f) sharpness map after image matting, and (g)–(h) ground-truth sharpness map.

does not work for flat image patches; therefore, this type of patches is avoided. Image matting [82] is used to fill up pixel positions belonging to flat patches, i.e., pixel positions that were not processed.

Sample results of the method proposed in [16] are presented in Fig. 2.17. The algorithm works well when the blur difference is considerable; however, its performance is poor for images with small amounts of blur.

2.5   REBLURRING-BASED METHOD

Without loss of generality, let us consider a 1D edge given by the expression $f(x) = Au(x) + B$ where A, B, and $u(x)$ represent the edge height, the edge offset, and the step function, respectively. Modeling the defocus blur as a convolution with a Gaussian Point Spread Function (PSF) with a unknown variance $k_0 = \mathcal{N}(0, \sigma_0)$, we have the blurred edge given by [114],

$$i(x) = f(x) * k_0. \tag{2.34}$$

The reblurred version of the edge with a second Gaussian kernel $k_r = \mathcal{N}(0, \sigma_r)$ is given by

$$i_r(x) = i(x) * k_r = f(x) * k_0 * k_r. \tag{2.35}$$

The expression for the gradient of the defocus blurred edge and the reblurred edge are given in Eqs. (2.34) and (2.35), respectively:

$$\nabla i(x) = \frac{A}{\sqrt{2\pi\sigma_0^2}} \exp\left(-\frac{x^2}{2\sigma_0^2}\right) \tag{2.36}$$

$$\nabla i_r(x) = \frac{A}{\sqrt{2\pi(\sigma_0^2 + \sigma_r^2)}} \exp\left(-\frac{x^2}{2(\sigma_0^2 + \sigma_r^2)}\right). \tag{2.37}$$

The quotient of Eqs. (2.34) and (2.35) is independent of the edge height,

$$Q = \frac{|\nabla i(x)|}{|\nabla i_r(x)|} = \sqrt{\frac{\sigma_0^2 + \sigma_r^2}{\sigma_0^2}} \exp\left(-\frac{x^2}{2\sigma_0^2} + \frac{x^2}{2(\sigma_0^2 + \sigma_r^2)}\right). \tag{2.38}$$

The maximum of Eq. (2.38) is reached at $x = 0$, i.e., at the edge position, it has the expression,

$$Q_{max} = \frac{|\nabla i(x = 0)|}{|\nabla i_r(x = 0)|} = \sqrt{\frac{\sigma_r^2 + \sigma_0^2}{\sigma_0^2}}. \tag{2.39}$$

Finally, the unknown defocus blur can be expressed as a function of the quotient of the magnitude gradients at the edge position as follows:

$$\sigma_0 = \frac{1}{\sqrt{Q_{max}^2 - 1}}\sigma_r. \tag{2.40}$$

With this approach, a sparse defocus blur map is created because the blur information is retrieved only at edge positions. The sparse blur information usually is propagated to other regions of the input image using matting Laplacian optimization [82]. Figure 2.18 presents results of the algorithm of Zhuo et al. [114]; the sparse blur map, as well as the whole, defocus blur after matting Laplacian are presented.

A similar approach to the one described above is presented in [109]. The sparse blur map is defined as explained above [114]; but, for the blur propagation, the image is segmented into superpixels [110], with the set of superpixels, a weighted connected graph is created. Then, the transductive inference method proposed by [111] is used to propagate the sparse blur information across the whole image. The proposed method is significantly faster than its predecessor [114].

Another method based on image reblurring is presented in [112]. The main difference lies in the definition of edge positions at different scales which is achieved using Canny edge detector [62] in a scale-space. The sparse blur map is defined only for edges that are consistent across the scale-space. This simple improvement provides the algorithm with robustness to noise. More variations of the principle presented in this section can be found in [45].

Figure 2.18: Sample results of the method proposed in [114] for several images of the dataset [38]. The first and last rows show the input images and the ground-truth blurring map, respectively. The second row presents the sparse blur map defined for edge positions in the input images; the third row shows the result of matting Laplacian optimization [82] applied to the sparse blur map.

In [40] a blur detection and classification method is proposed. The blur is detected in a coarse to fine iterative approach. A coarse version of the blurring map is first defined based on two facts. First, the observation that the second convolution, a.k.a. reblurring, of an image by the same PSF produces a marginal degradation, and second, the average power spectrum of natural images follows a distribution given by $EY(f) \propto 1/f$.

Given an input image y and its corresponding Fourier transform as $Y = \mathcal{F}\{y\}$, and creating a Gaussian reblurring kernel k_r and its corresponding optical transfer function (OTF) $K_r = \mathcal{F}\{k_r\}$, the coarse spectrum is defined as follows:

$$B = \mathcal{F}^{-1}\{\log(\text{real}(Y)) - K_r \cdot \log(\text{real}(Y))\}. \tag{2.41}$$

The first step in the iterative update consists of a segmentation using superpixels approach of [110]. The average blur in each region is used as the initial blurriness value, this value is updated considering color and gradient distribution similarity of the regions and its neighboring regions.

CIE LAB color space is used in the color similarity. Each region is assigned as color descriptor c the mean color of the pixels forming the region. Then, the color similarity cs_{ij} between two neighboring regions c_i, c_j $i \neq j$ is defined using the Euclidean distance $d(c_i, c_j)$ as

$$gs_{ij} = \begin{cases} d(c_i, c_j) \text{ as } cs_{ij} = e^{-\frac{d(c_i, c_j)}{\sigma_1^2}}, & \text{if } j \in \Omega_i \\ 0, & \text{otherwise,} \end{cases} \qquad (2.42)$$

where $\sigma_1^2 = 0.1$ and Ω_i is the set of neighbor regions of i. For the gradient distribution similarity, the gradient covariance matrix distance is used. The affine-invariant distance metric between the covariance matrices of two neighboring regions C_i, C_j is defined using the matrix Frobenius norm $\|\cdot\|_F$ as follows:

$$d_{AI}(C_i, C_j) = \left\| \log \left(C_i^{-\frac{1}{2}} C_j C_i^{-\frac{1}{2}} \right) \right\|_F. \qquad (2.43)$$

The gradient distribution similarity is given by

$$gs_{ij} = \begin{cases} e^{-\frac{d_{AI}(C_i, C_j)}{\sigma_2^2}}, & \text{if } j \in \Omega_i \\ 0, & \text{otherwise,} \end{cases} \qquad (2.44)$$

where the similarity s_{ij} between two regions is given by $s_{ij} = \beta \cdot cs_{ij} + (1 - \beta) \cdot gs_{ij}$ with $\beta = 0.4$ in order to emphasize the gradient similarity, and $\sigma_2^2 = 0.1$.

Finally, the blur map is iteratively updated. The update process is developed for each image region individually and it considers the influence of neighboring regions as follows:

$$B_i^{t+1} = coh_i \cdot B_i^t + (1 - coh_i) \cdot \sum_{q=1}^{N} s_{ij} \cdot b_q^t, \qquad (2.45)$$

where $coh_i = (1 + \max s_{ij})^{-1}$ is the measure of coherence of region i with respect to its current blurriness, N is the total number of image regions in Ω_i. Figure 2.19 presents results of the method proposed in [40] for several images of the dataset [38]. The negative effect of the superpixel-based segmentation is clearly perceivable.

2.6   DESCRIPTION OF THE PROPOSED METHOD

We propose a new local blur metric defined as the quotient of the mean of high-frequency coefficients to the mean of low-frequency coefficients of the log-spectrum of the image gradients. To avoid the use of propagation algorithms [82] which are commonly used when only sparse blur maps are available [22, 113, 114], we compute the aforementioned quotient of bands for all the image patches. We also propose a simple yet effective method to segment out-of-focus regions using a global threshold which is defined using weak textured regions present in the

Figure 2.19: Sample results of the method proposed in [40] for several images of the database [38]. The first column shows the input images, the second row the results of [40], and the third column the ground-truth blur maps provided by [38].

input image. Our method was evaluated using two publicly available datasets. The results show that the proposed algorithm outperforms state-of-the-art methods both, qualitatively as well as quantitatively.

Figure 2.20 shows the block diagram of the proposed method. Although the spectral power is not isotropic in every direction, it has been proved to be larger at horizontal and vertical directions [73]; therefore, using the luminance component of the input image, the horizontal and vertical gradient images are computed as G_x and G_y, respectively. The computed gradient images are used to generate a spatially varying sharpness map based on the log spectrum of local gradient image patches followed by edge-preserving filtering and texture-based thresholding. Details about the sharpness map generation, edge-preserving filtering, and texture-based thresholding are presented in the subsequent sections.

2.6.1   PROPOSED SHARPNESS METRIC

Most state-of-the-art methods devoted to defining a blur map involve the use of spectral information [17, 34, 37, 40, 113, 117]. Here we propose the use of the DFT over fully overlapping patches of the gradient images.

It is known that the 2D power spectrum of natural images varies considerably between different images and within an image when considering different orientations. The aforementioned 2D power spectrum can be reduced to a 1D function of spatial frequency by averaging

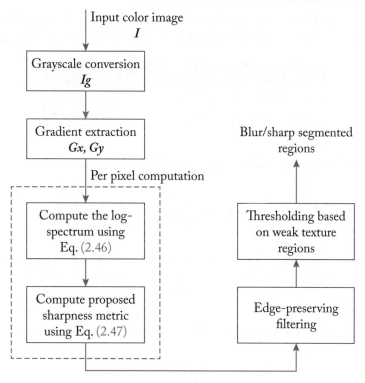

Figure 2.20: Flowchart of the proposed algorithm [115].

across all the orientations [73]. Empirically, it has been shown that the spectral power, averaged over several natural images, falls with the spatial frequency according to $1/f^{\alpha}$ where α is close to 2 [73]. There are several theories that attempt to explain this behavior. One of the most accepted beliefs establishes that it is due to scale invariance of the visual world [73]; another theory says that this behavior of the power spectrum is due to the existence of edges since edges have a $1/f^2$ power spectrum [118]. We have estimated the average power spectrum of the 200 testing images of the BSDS500 Berkeley segmentation dataset [116] using both the luminance and the image gradients; average results for the slope of the power spectrum using linear regression are presented in Table 2.1. The slope of the power spectrum of the image gradients is more stable than the power spectrum computed using the image luminance. Motivated by this fact, we propose the use of the log-power spectrum of the image gradients [113]. Therefore, for each image pixel, we define $N \times N$ patches of the input image gradients in the horizontal and vertical directions, G_x and G_y, respectively. We define the log-power spectrum of the $N \times N$ gradient patch P_{G_k},

Table 2.1: Average power spectrum parameters for the 200 testing images of the BSDS500 Berkeley segmentation dataset [116]. Both luminance and image gradients are evaluated.

Input	Slope mean	Slope variance
Luminance	2.4	0.069
Gradients	1.5	0.006

$k \in \{x, y\}$, centered at the pixel position (i, j) as follows:

$$F_{k(i,j)}(\omega_1, \omega_2) = \log_{10} \left(\left(\frac{|\mathcal{F}(P_{G_k}(i,j))|}{N^2} \right)^2 \right). \tag{2.46}$$

In Eq. (2.46), the gradient patch P_{G_k} has its mean value removed and $\mathcal{F}(\cdot)$ represents the Fast Fourier Transform (FFT). Having a power spectrum that falls off as $1/f^2$ means that equal energy will be in equal octaves [73]; this added to the fact that out-of-focus blur tends to reduce the energy in high-frequency bands [37] motivate us to propose a sharpness metric defined as the quotient of the mean of high-frequency bands to the mean of the low-frequency bands of the average log-power spectrum as follows:

$$S_k(i, j) = \frac{\overline{F_{k(i,j)}^{HP}}}{\overline{F_{k(i,j)}^{LP}}} \quad k \in \{x, y\}, \tag{2.47}$$

where $\overline{F_{k(i,j)}^{LP}}$ and $\overline{F_{k(i,j)}^{HP}}$ are the mean of the low- and high-frequency bands of $F_{k(i,j)}(\omega_1, \omega_2)$, respectively. The low-frequency $F_{k(i,j)}^{LP}$ and high-frequency $F_{k(i,j)}^{HP}$ bands of $F_{k(i,j)}(\omega_1, \omega_2)$ are defined as follows:

$$F_{k(i,j)}^{LP} = F_{k(i,j)}(\omega_1, \omega_2) \quad \text{for} \quad \sqrt{\omega_1^2 + \omega_2^2} \le \frac{\pi}{4} \tag{2.48}$$

$$F_{k(i,j)}^{HP} = F_{k(i,j)}(\omega_1, \omega_2) \quad \text{for} \quad \sqrt{\omega_1^2 + \omega_2^2} \ge \frac{\pi}{2}. \tag{2.49}$$

Based on performance tests, we decided not to include the range of frequencies $\frac{\pi}{4} < \sqrt{(\omega_1^2 + \omega_2^2)} < \frac{\pi}{2}$. A dense sharpness map is formed with the sharpness metric S_k defined in Eq. (2.47) as follows:

$$M_o(i, j) = S_x^2(i, j) + S_y^2(i, j). \tag{2.50}$$

In Fig. 2.21 an illustration of the proposed sharpness metric given by Eq. (2.50) for the images *out_of_focus0250.jpg* and *out_of_focus0305.jpg* of the publicly available dataset [38] is presented; smaller values of the proposed sharpness metric correspond to sharper patches on the testing images.

Figure 2.21: Illustration of the proposed sharpness metric given by Eq. (2.50) for two images of the publicly available dataset [38]. Smaller values of the proposed sharpness metric correspond to sharper patches.

2.6.2   EDGE-PRESERVING FILTERING

In order to homogenize the preliminary sharpness map M_o within neighboring areas sharing similar sharpness conditions but, keeping at the same time the boundaries between blur and sharp areas, the map M_o is smoothed using an edge-preserving guided filter approach [119]. The output of the guided filter for pixel (i, j) is given by:

$$M_f(u, v) = a(i, j)I_g(u, v) + b(i, j) \quad \forall (u, v) \in \omega(i, j), \tag{2.51}$$

where I_g is the filtering guidance map which here corresponds to the input image, $\omega(i, j)$ is a window of size $r \times r$ centered at pixel (i, j), and $a(i, j), b(i, j)$ are constant coefficients within $\omega(i, j)$. This model ensures that the filtered image will have an edge only where the guidance map I_g has one because $\nabla M_f = a \nabla I_g$, where ∇ is the gradient operator. The coefficients $a(i, j)$, $b(i, j)$ are obtained for each window $w(i, j)$ by minimizing the following cost function [119]:

$$C(i, j) = \sum_{(u,v) \in \omega(i,j)} \left[D(i, j)^2 + \lambda a(i, j)^2 \right], \tag{2.52}$$

where the data fitting term $D(i, j)$ is defined as $D(i, j) = a(i, j)G(u, v) + b(i, j) - M_o(u, v)$ and λ is a regularization constant that penalizes large values of $a(i, j)$. In our implementation and for all the results reported here we use $r = 15$ and $\lambda = 1 \times 10^{-6}$. The optimal solutions for $a(i, j)$ and $b(i, j)$ in Eq. (2.52) are given by [119]:

$$a(i, j) = \frac{\sum_{(u,v) \in \omega(i,j)} I_g(u, v)M_o(u, v) - \mu(i, j)\overline{M_o}(u, v)}{[\omega(i, j)] (\sigma(i, j)^2 + \lambda)} \tag{2.53}$$

$$b(i, j) = \overline{M_o}(i, j) - a(i, j)\mu(i, j), \tag{2.54}$$

where $[\omega(i, j)]$ represents the number of elements in $\omega(i, j)$, $\overline{M_o}(i, j)$ is the mean of M_o within the window $\omega(i, j)$ and $\mu(i, j)$ and $\sigma(i, j)^2$ are the mean and variance of I_g within the window $\omega(i, j)$, respectively.

Let $\psi(i, j)$ be the set containing the indices of all overlapping windows containing the pixel (i, j). In (2.53) and (2.54), it can be seen that a pixel (i, j) contributes to determining all $a(u, v)$ and $b(u, v)$ with $(u, v) \in \psi(i, j)$, then the final filtered output is given by:

$$M_f(i, j) = \frac{1}{[\omega]} \sum_{(u,v) \in \psi(i,j)} a(u, v) I_g(i, j) + b(u, v). \tag{2.55}$$

The preliminary sharpness map M_o as well as the filtered output M_f for three images of the dataset [38] are shown in rows (b) and (c) of Fig. 2.22, respectively.

2.6.3 FOREGROUND-BACKGROUND SEPARATION BASED ON WEAK TEXTURED REGIONS

Segmentation of the sharpness map into foreground and background region is a difficult task; several authors do not include this final step in their work [16, 17, 40, 114], and those who include it most of the time trust in the matting Laplacian optimization [82]. However, solutions of an alpha matting algorithm are extremely sensitive to the initialization. Therefore, we propose a segmentation based on a single threshold defined for each image.

Edge detection is associated with abrupt changes of intensity, i.e., high-frequency content; therefore, edges detected in an image that includes out-of-focus blur, most of the time are within the sharp region and the boundaries between sharp and blur regions. Based on this observation, we use the positions of pixels that belong to edges in the input image and M_f to create an enhanced map M_{fe} which later is thresholded using a global threshold Th.

In order to define M_{fe} we first detect edges in the input image using Sobel edge detector. Let E be the number of pixels detected as edges, then with the pixel positions e_k for $k \in [1, \ldots, E]$ we define the enhanced map as follows:

$$M_{fe}(i, j) = \log_{10}\left(1 + |M_f(i, j)| \cdot F(i, j)\right), \tag{2.56}$$

where $F(i, j)$ is the cardinality of the set defined for every pixel position (i, j) as follows:

$$\{1 \mid M_f(i, j) < M_f(e_k)\}, \quad \forall k \in [1, \ldots, E]. \tag{2.57}$$

Since defocus blur removes or at least weakens the image texture; in order to define the global threshold Th we first define the set Γ as formed for the positions of all pixels that belong to weak textured regions, then Th is defined as follows:

$$Th = \max\left\{M_{fe}(i, j)\right\}, \quad \forall(i, j) \in \Gamma. \tag{2.58}$$

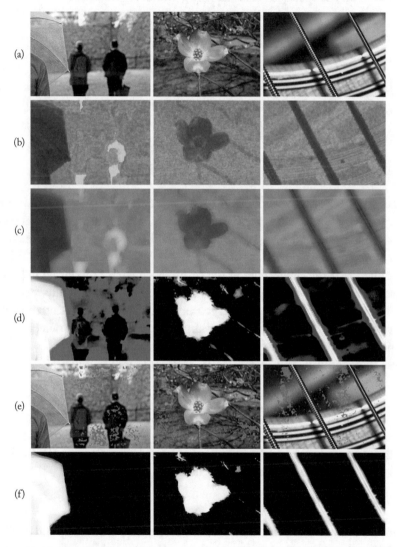

Figure 2.22: Sample results of different stages of the proposed method. (a) Input images from the database [38], (b) preliminary sharpness map M_o defined by Eq. (2.50), (c) sharpness map after guided edge preserving filtering as defined by Eq. (2.55), (d) enhanced sharpness map as defined by Eq. (2.56), (e) detected weak texture regions colored in red, and (f) final segmented sharpness map as defined by Eq. (2.61).

To define Γ, i.e., the set of positions of weak texture regions caused by defocus blur, we first create a map based on the spectrum contrast [113], which is given by:

$$T_s(i, j) = \sum_{\gamma \in \{r,c\}} \left| \frac{1}{e^{\phi_\gamma(i,j)} - \sigma_b^2} \right|, \tag{2.59}$$

where $\phi_\gamma(i, j)$ is the inverse Fourier transform of $\Phi_\gamma(\omega_1, \omega_2)$, Φ_r, and Φ_c are the spectrum contrast along rows and columns, respectively, which are defined as:

$$\Phi_\gamma(\omega_1, \omega_2) = \left| \log |I_f(\omega_1, \omega_2)| - \frac{1}{Q} \sum_{(\omega_x, \omega_y) \in \Omega_\gamma} \log |I_f(\omega_x, \omega_y)| \right|. \tag{2.60}$$

In Eq. (2.60) to reduce the influence of image noise, I_f is defined as the FT of the input image after being filtered with a Gaussian kernel with $\sigma_b = 0.5$; Ω_r and Ω_c are the vertical and horizontal windows centered at position (ω_1, ω_2) of sizes $1 \times Q$ and $Q \times 1$, respectively. In our implementation and for all the results reported we used $Q = 23$. Weak texture regions on the input image are defined by analyzing the eigenvalues of the covariance matrix as reported in [120, 121]. The method proposed in [120] was adopted and applied to $T_s(i, j)$. Naturally textureless regions that belong to the focused region may be flagged as weak texture regions caused by defocus blur, e.g., small parts of leaves, flowers, and human skin; therefore, we discard detected spots of weak texture regions having a connectivity smaller than 15 pixels.

The final segmented sharpness map $M_s(i, j)$ is computed as follows:

$$M_s(i, j) = \begin{cases} M_{fe}(i, j), & \text{if } M_{fe}(i, j) \geq Th. \\ 0, & \text{if } M_{fe}(i, j) < Th, \end{cases} \tag{2.61}$$

where M_{fe} is the enhanced map as defined in Eq. (2.56). Additionally, in order to obtain a binary map $M_s(i, j)$, $M_{fe}(i, j)$ can be replaced with 1 in Eq. (2.61).

2.6.4   RESULTS

In this section, we evaluate the performance of the method quantitatively and qualitatively. For quantitative comparison, we evaluate our method using the public dataset of Shi et al. [38] which provides 704 images partially affected by defocus blur. We use the precision-recall curves in order to compare quantitatively the performance of the proposed algorithm. The precision is the number of pixels correctly selected divided by the total number of selected pixels. The recall, on the other hand, is the number of pixels that were selected correctly divided by the total number of ground truth pixels. The aforementioned description for precision and recall are defined by Eqs. (2.62) and (2.63) where B_D represents the set of detected blurred pixels and B_{GT} represents the set of ground truth blurred pixels, for ground truth blurred pixels we use the

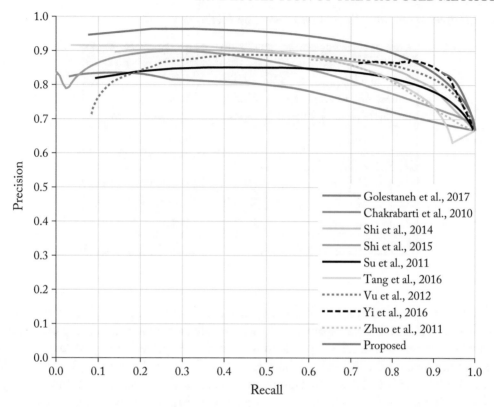

Figure 2.23: Precision-recall curves for [16, 17, 25, 34, 37, 39, 40, 114, 117] and the proposed method evaluated using the defocus images from the database [38]. The proposed method achieves the highest precision when recall is larger than 0.91. Note that low values of recall correspond to poor detection of the correct pixels so a high precision with a low recall can be reached by detecting correctly a single pixel.

hand-segmented blurred pixels provided in the dataset [38]:

$$precision = \frac{B_D \cap B_{GT}}{B_D} \qquad (2.62)$$

$$recall = \frac{B_D \cap B_{GT}}{B_{GT}}. \qquad (2.63)$$

Figure 2.23 shows precision-recall curves for the proposed method together with nine existing methods including Golestaneh et al. [17], Chakrabarti et al. [34], Shi et al., [37], Shi et al. [16], Su et al. [39], Tang et al. [40], Vu et al. [117], Yi et al. [25], and Zhuo et al. [114]. The code made publicly available by these authors was used in all cases. These results show that the

Table 2.2: Average running time of blur detection methods

Method	Time [s]	Implementation
Vu et al. [113]	13.09	Matlab
Golestaneh et al. [17]	213.5	Matlab
Chakrabarti et al. [34]	3.36	Matlab
Shi et al. [16]	14.53	Matlab and C mex
Su et al. [39]	16.32	Matlab
Tang et al. [40]	0.055	Matlab and C mex
Yi et al. [25]	29.44	Matlab and C mex
Zhuo et al. [110]	21.21	Matlab
Proposed	32.71	Matlab

proposed method outperforms the competing algorithms for the higher values of recall which is the goal in a precision–recall space [122]. Additionally, a running time comparison for the evaluated blur detection methods is provided in Table 2.2. Reported timing corresponds to average running time on the 704 out-of-focus images provided in the dataset [38]. An Intel(R) Core(TM) i7-5500 CPU @2.4 GHz laptop with 16 GB of RAM installed was used. The proposed algorithm is more than six times faster on average than the closest competing algorithm in the precision–recall curve, i.e., Golestaneh et al. [17]. For qualitative comparison, Fig. 2.24 shows results in several images of the dataset [38]. Clearly, the proposed algorithm provides results closer to the ground truth; moreover, only the proposed algorithm and Yi et al. [25] can provide background/foreground segmented results, i.e., binary masks besides the classic blur map.

In Vu et al. [117], subjective sharpness maps were created for six color images. The judgment of 11 subjects was averaged to create subjective sharpness maps with a resolution of 8×8 pixels which have at least 206 and at most 245 different sharpness levels. Qualitative results on the dataset [117] are presented in Fig. 2.25. Since in this case the ground truth sharpness map is not binary as in [38], we have included not only the result given by M_s but also the map before global thresholding, i.e., M_{fe} given by Eq. (2.56). In order to compare the similarity between the grayscale ground-truth sharpness maps provided in [117] and the results of blur detection methods, we used the full reference structural similarity index (SSIM) [123]. Comparative results of the SSIM index are presented in Fig. 2.26 where the proposed method shows better performance.

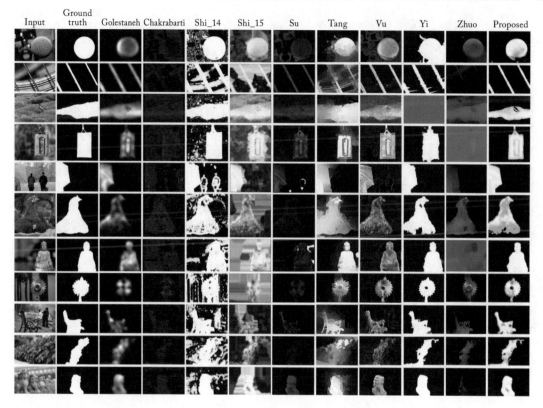

Figure 2.24: Blur detection results achieved by the methods [16, 17, 25, 34, 37, 39, 40, 114, 117] and the proposed one over images of the publicly available database [38].

2.6.5   CONCLUSION

In this chapter we have presented a novel algorithm to detect defocus blur from a single image. The proposed method uses a blur/sharpness metric defined as the quotient of high- to low-frequency bands of the log-spectrum of the image gradients. Additionally, a foreground-background segmentation that uses a global threshold defined using weak textured regions on the input image is presented. The proposed segmentation approach avoids the commonly used alpha matting algorithm which is very sensitive to the initialization. The proposed method outperforms state-of-the-art methods both quantitatively and qualitatively and its non-optimized Matlab implementation shows a competitive running time.

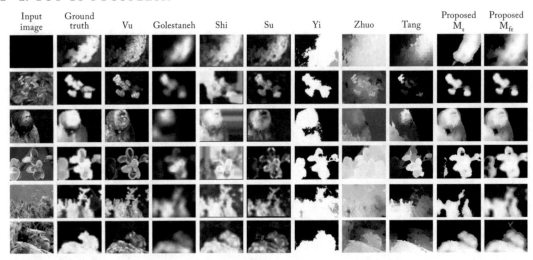

| Input image | Ground truth | Vu | Golestaneh | Shi | Su | Yi | Zhuo | Tang | Proposed M_s | Proposed M_{fe} |

Figure 2.25: Blur detection results achieved by the methods [16, 17, 25, 39, 40, 114, 117] and the proposed one over images of the publicly available database [42]. The proposed method, before and after global thresholding, i.e., M_{fe} and M_s, respectively, are included for comparison because the ground-truth sharpness map in the evaluated database is not binary.

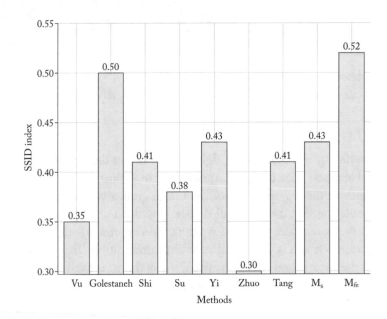

Figure 2.26: SSID index between the ground-truth sharpness map provided by the publicly available database [42] and the blur detection results of the methods [16, 17, 25, 34, 37, 39, 40, 114, 117] and the proposed one with and without thresholding, M_s and M_{fe}, respectively.

2.7   SUMMARY

This chapter covered the most relevant methods used to estimate a sharpness map from a single image. A variety of techniques have been presented and their performance evaluated. Most of the state-of-the-art methods for sharpness map estimation use a combination of several metrics and usually multi-scale approaches. Additionally, a novel sharpness map estimation based on the quotient of the mean of high-frequency coefficients to the mean of low-frequency coefficients of the log-spectrum of the image gradients is presented. We also proposed a simple yet effective method to segment out-of-focus regions using a global threshold which is defined using weak textured regions present in the input image. Our method was evaluated using two publicly available datasets.

CHAPTER 3

Image Quality Assessment

Digital images and videos are produced every day in astonishing amounts and the demand for higher quality is constantly rising which creates a need for advanced image quality assessment. Additionally, image quality assessment is important for the performance of image processing algorithms. It has been determined that image noise and artifacts can affect the performance of algorithms such as face detection and recognition [125], image saliency detection [126], and video target tracking [127]. Therefore, image quality assessment (IQA) has been a topic of intense research in the fields of image processing and computer vision. Since humans are the end consumer of multimedia signals, subjective quality metrics provide the most reliable results; however, their cost in addition to time requirements makes them unfeasible for practical applications; thus, OQM are usually preferred.

In this chapter, we first discuss and evaluate relevant work for image sharpness assessment. The principles behind Full-reference (FR), reduced-reference (RR), and no-reference (NR) methods are presented. Later, a comprehensive analysis of no-reference image sharpness assessment (NRSA) methods is provided. Finally, a novel approach for NRSA based on perceptually weighted image gradients is presented. The proposed NRSA method is evaluated over six subject-rated publicly available databases. Results show that the proposed method correlates well with perceived sharpness and provides competitive performance when compared with state-of-the-art algorithms.

3.1 FULL-REFERENCE IMAGE QUALITY ASSESSMENT

Full-reference image quality assessment (FR-IQA) methods [123, 124, 128, 130–132, 134, 149, 150] provide the best objective results; however, the need of the ground truth pristine image makes this approach unsuitable in most practical applications. FR-IQA approaches such as the SNR, peak signal-to-noise ratio (PSNR), and mean squared error (MSE) are attractive not only for their low computational burden but also by their clear physical meaning as well as their convexity and differentiability properties [135]; however, it is widely acknowledged that they do not correlate well with subjective judgments of image quality [135]. Initially, FR-IQA methods made an emphasis on modeling the effects of contrast and luminance [136]. Since human visual perception is highly specialized for extracting structural information Wang et al. [123] present an algorithm that exploits this fact. A multi-scale and an information-weighted version of [123] are presented by Wang et al. in [132] and [131], respectively. Sheikh et al. [137] propose a parameterless algorithm that uses an information fidelity criterion based on natural image statistics; this approach is further improved in [130] by linking the distortion process to the loss of image information. The authors of [138] use the first- and second-order Riesz-transform coefficients to tailor image features, the same authors propose in [139] the spectral residual-based similarity (SR-SIM) which assumes that the saliency map of an image is closely related to its perceived quality. Gradient similarity together with luminance and contrast changes are incorporated in the method proposed by Liu et al. [149]. A method that exploits the similarity of the gradient magnitude is proposed by Xue et al. [150]. The authors of [140] combine natural image statistics with multi-resolution methods to estimate the image quality. Zhang et al. [141] propose a method based on visual saliency.

3.2   MATHEMATICALLY DEFINED IMAGE QUALITY METRICS

The first step in OQM was based on numerical measures evaluating luminance differences between corresponding pixels in the original and degraded images. These methods were conceived for the evaluation of image compression algorithms, However, although some numerical measures correlate well with human perception for a given compression algorithm, they are not reliable for evaluation across different compression methods [142]. Here we include the most used:

Average difference:

$$AD = \sum_{i=1}^{M} \sum_{j=1}^{N} \frac{X(i,j) - Y(i,j)}{MN}. \tag{3.1}$$

Average difference:

$$SC = \frac{\sum_{i=1}^{M} \sum_{j=1}^{N} X^2(i,j)}{\sum_{i=1}^{M} \sum_{j=1}^{N} Y^2(i,j)}. \tag{3.2}$$

Normalized cross-correlation:

$$NCC = \frac{\sum_{i=1}^{M} \sum_{j=1}^{N} X(i,j)Y(i,j)}{\sum_{i=1}^{M} \sum_{j=1}^{N} X^2(i,j)}. \tag{3.3}$$

Correlation quality:

$$CC = \frac{\sum_{i=1}^{M} \sum_{j=1}^{N} X(i,j)Y(i,j)}{\sum_{i=1}^{M} \sum_{j=1}^{N} X(i,j)}. \tag{3.4}$$

Maximum difference:

$$MD = \max\left(|X(i,j) - Y(i,j)|\right). \tag{3.5}$$

Image fidelity:

$$IF = 1 - \frac{\sum_{i=1}^{M} \sum_{j=1}^{N} (X(i,j) - Y(i,j))}{\sum_{i=1}^{M} \sum_{j=1}^{N} X^2(i,j)}. \qquad (3.6)$$

Laplacian mean square error:

$$LMSE = \frac{\sum_{i=1}^{M} \sum_{j=2}^{N} (\Phi_L [X(i,j)] - \Phi_L [Y(i,j)])^2}{\sum_{i=1}^{M} \sum_{j=2}^{N} (\Phi_L [X(i,j)])^2}. \qquad (3.7)$$

Peak mean square error:

$$PMSE = \frac{1}{MN} \frac{\sum_{i=1}^{M} \sum_{j=1}^{N} (X(i,j) - Y(i,j))}{\max (X(i,j))^2}. \qquad (3.8)$$

Normalized absolute error:

$$NAE = \frac{\sum_{i=1}^{M} \sum_{j=1}^{N} |X(i,j) - Y(i,j)|}{\sum_{i=1}^{M} \sum_{j=1}^{N} |X(i,j)|}. \qquad (3.9)$$

Normalized mean square error:

$$NMSE = \frac{\sum_{i=1}^{M} \sum_{j=1}^{N} (X(i,j) - Y(i,j))^2}{\sum_{i=1}^{M} \sum_{j=1}^{N} (X(i,j))^2}. \qquad (3.10)$$

L_p-norm:

$$L_p = \left[\frac{1}{MN} \sum_{i=1}^{M} \sum_{j=1}^{N} |X(i,j) - Y(i,j)|^p \right]^{1/p} \qquad p = 1, 2, 3, \qquad (3.11)$$

where X, Y are the original and degraded images, respectively. The images have dimensions $M \times N$ pixels, and

$$\Phi_L[X(i,j)] = X(i+1,j) + X(i-1,j) + X(i,j+1) + X(i,j-1) - 4X(i,j). \qquad (3.12)$$

Several FR-IQA methods have been proposed based on the aforementioned expressions. In [143] the applicability of different contrast sensitivity functions (CSF) in the quality assessment of half-toning images is studied. Similar to [144], a low-pass CSF is shown to provide better results than a bandpass CSF; therefore, a CSF weighted MSE is proposed. In [145] the new weighted Mean Square Error (NwMSE) is presented. The variance of the image block around each pixel is used as a per-pixel weighting factor. In [146] the reference and testing images are first transformed to the contrast domain and then the MSE is defined as a quality metric. Improvements to the SNR and PSNR indexes which include analysis in the frequency domain and HVS clues are proposed in [147, 148].

3.3 STRUCTURAL-BASED APPROACHES

A Universal Image Quality Index (UQI)

In [136], image distortion is modeled as the combination of three factors: loss of correlation, luminance distortion, and contrast distortion:

$$Q = \frac{\sigma_{xy}}{\sigma_x \sigma_y} \cdot \frac{2\bar{X}\bar{Y}}{\bar{X}^2 + \bar{Y}^2} \cdot \frac{2\sigma_x \sigma_y}{\sigma_x^2 + \sigma_y^2}, \quad (3.13)$$

where

$$\bar{X} = \frac{1}{MN} \sum_{i=1}^{M} \sum_{j=1}^{N} X(i,j), \quad \bar{Y} = \frac{1}{MN} \sum_{i=1}^{M} \sum_{j=1}^{N} Y(i,j)$$

$$\sigma_x^2 = \frac{1}{(M-1)(N-1)} \sum_{i=1}^{M} \sum_{j=1}^{N} (X(i,j) - \bar{X}), \quad \sigma_y^2 = \frac{1}{(M-1)(N-1)} \sum_{i=1}^{M} \sum_{j=1}^{N} (Y(i,j) - \bar{Y})$$

$$\sigma_{xy} = \frac{1}{(M-1)(N-1)} \sum_{i=1}^{M} \sum_{j=1}^{N} (x(i,j) - \bar{X})(y(i,j) - \bar{Y}).$$

$$(3.14)$$

In Eq. (3.13) the left term represents the correlation between X and Y, and the second and third represent the luminance and contrast similarity between corresponding positions in the original and degraded images, respectively. Since image quality is space variant, Eq. (3.13) is applied to full overlapping patches of $B \times B$ pixels in the testing images. The final image quality index is defined as follows:

$$Q_{UQI} = \frac{1}{M} \sum_{j=1}^{M} Q_j, \quad (3.15)$$

where M is the total number of overlapping patches on the image under evaluation and Q_j is the jth image patch.

Image Quality Assessment Based on Gradient Similarity

In [129] the image quality index is defined as follows:

$$GS = \frac{2\mathbf{G}_x \cdot \mathbf{G}_y + c}{2\mathbf{G}_x^2 + \mathbf{G}_y^2 + c}, \quad (3.16)$$

where $\mathbf{G}_x$ and $\mathbf{G}_y$ are defined for each pixel as the maximum value of the gradient computed with four different kernels:

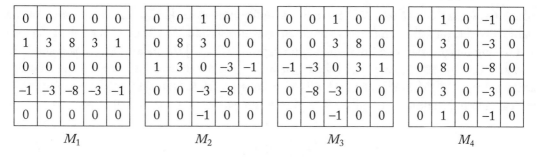

Figure 3.1: Kernels used to compute gradients $\mathbf{G}_x$ and $\mathbf{G}_y$ in [129].

$$\mathbf{G}_x = \max_{k=1,2,3,4} mean2|\cdot M_k|, \tag{3.17}$$

where c is a constant introduced for numerical stability, $mean2$ represents the mean value of a matrix, and M_k is the gradient kernel as shown in Fig. 3.1.

Gradient magnitude similarity deviation (GMSD)

In [133] a pixel-wise gradient magnitude similarity (GMS) between the reference and the testing images is defined as follows:

$$GMS = \frac{2\mathbf{G}_x \cdot \mathbf{G}_y + c}{2\mathbf{G}_x^2 + \mathbf{G}_y^2 + c}, \tag{3.18}$$

where c is a constant introduced for numerical stability and $\mathbf{G}_x$ and $\mathbf{G}_y$ represent the gradient magnitude of the reference and testing images, respectively. The gradient magnitudes are computed using Prewitt filters [151]. Finally, the GMSD index is defined as the standard deviation of GMS.

3.4   STRUCTURAL SIMILARITY

Structural similarity index (SSIM)

Based on the fact that HVS is highly adapted to extract structure information from the viewing field, the idea of using structure distortion as an approximation to perceived image distortion is proposed [152]. This idea has been highly exploited in several publications [131, 132, 136, 139, 153].

Since the luminance of an object is the product of the illumination and the reflectance, in order to use the structural information of an image, the illumination influence should be removed. In [136], a similarity measure involving three components is proposed as follows:

$$SSIM = \frac{2\mu_x\mu_y + C_1}{\mu_x^2 + \mu_y^2 + C_1} \cdot \frac{2\sigma_x\sigma_y + C_2}{\sigma_x^2 + \sigma_y^2 + C_2} \cdot \frac{2\sigma_{xy} + C_2}{2\sigma_x\sigma_y + C_2}, \tag{3.19}$$

where C_1, C_2 are constants used to avoid instability present in [153]. The three terms in the right side of Eq. (3.19) compare the luminance, contrast, and structure between the reference and testing images. Although in similar approaches [152, 153] the means μ_x, μ_y and standard deviations σ_x, σ_y, and σ_{xy} are locally defined using 8×8 pixels blocks, in order to avoid blocking artifacts in the SSIM index map a 11×11 circular-symmetric Gaussian weighting function w with standard deviation of 1.5 pixels is used to estimated local statistics as follows:

$$\mu_x = \sum_{i=1}^{N} \sum_{j=1}^{N} w(i,j) X(i,j) \tag{3.20}$$

$$\sigma_x = \sqrt{\sum_{i=1}^{N} \sum_{j=1}^{N} w(i,j) \left(X(i,j) - \mu_x \right)^2} \tag{3.21}$$

$$\sigma_{xy} = \sum_{i=1}^{N} \sum_{j=1}^{N} w(i,j) \left(X(i,j) - \mu_x \right) \left(Y(i,j) - \mu_y \right). \tag{3.22}$$

Finally, the mean of the SSIM index map (MSSIM) is used as a global image quality metric.

A fast and high-performance IQA index based on spectral residual (SR-SIM)

Based on the hypothesis that an image's visual saliency map is closely related to its perceived quality, in [139] an IQA method that exploits spectral residual visual saliency (SRVS) and local contrast is proposed. For the reference and testing images the SRVS, i.e., R_x and R_y, are defined as follows:

$$R_x(u,v) = \log |\mathcal{F}X(i,j)| - h_n(u,v) \otimes \log |\mathcal{F}X(i,j)|, \tag{3.23}$$

where $\mathcal{F}$ represents the FT, and $h_n(u,v)$ is an $N \times N$ averaging filter.

For the reference and testing images, the local contrast is computed as follows:

$$G_x(i,j) = \sqrt{G_{xx}^2 + G_{yy}^2}, \tag{3.24}$$

where the gradients of X, i.e., G_{xx} and G_{yy} are computed using a $[3 \times 3]$ Scharr gradient operator [154]. The local similarity map between the reference and testing images X, Y, respectively, are defined as follows:

$$S(i,j) = \left[\frac{2R_x R_y + C_1}{R_x^2 + R_y^2 + C_1} \right] \cdot \left[\frac{2G_x G_y + C_2}{G_x^2 + G_y^2 + C_2} \right]^{\alpha}, \tag{3.25}$$

where the right-hand terms in Eq. (3.25) correspond to the spectral residual visual saliency and the local contrast between both images, and α is a constant used to adjust the relative importance of both terms. Finally, the overall similarity between both images is defined as follows:

$$SR - SIM = \frac{\sum_{(i,j)\in\Omega} S(i,j) \cdot R_m(i,j)}{\sum_{\forall(i,j)} R_m(i,j)},$$ (3.26)

where Ω is the whole image spatial domain, $R_m(i,j)$ is the higher visual salience found in both images, i.e, $R_m(i,j) = \max\{R_x(i,j), R_y(i,j)\}$.

Multi-scale structural similarity for image quality assessment (MS-SIM)

In [153] the SSIM was introduced. The same authors provided an improvement in [136] by introducing a set of constants that avoids numerical instability. The general form of SSIM is given by

$$SSIM = [l(X,Y)]^\alpha \cdot [c(X,Y)]^\beta \cdot [s(X,Y)]^\gamma,$$ (3.27)

where α, β, and γ are constants to set the relative importance among the luminance term $l(X,Y)$, the contrast term $c(X,Y)$, and the structure term $s(X,Y)$ which are computed using the reference and testing images X, Y, respectively. In [132], the method of [136] is extended to a multi-scale framework. The proposed system includes M scales, for every scale the contrast and similarity terms are computed, and the luminance term is only computed at the scale M, the multi-scale SSIM expression is as follows:

$$MS - SSIM = [l_M(X,Y)]^{\alpha M} \cdot \prod_{j=1}^{M} [c_j(X,Y)]^{\beta j} \cdot [s_j(X,Y)]^{\gamma j}.$$ (3.28)

Five scales are reported in [132]. The parameters in Eq. (3.28) are defined using a subjective test that involved eight subjects. The defined parameters are: $\beta_1 = \gamma_1 = 0.0488$, $\beta_2 = \gamma_2 = 0.2856$, $\beta_3 = \gamma_3 = 0.3001$, $\beta_4 = \gamma_4 = 0.2363$, and $\alpha_5 = \beta_5 = \gamma_5 = 0.1333$.

A feature-based image quality assessment metric using Riesz transform (RFSIM)

It has been shown that the HVS perceives an image based on its low-level features such as edges [156] and corners [155]. Therefore, in [138] a Riesz-transform based feature similarity index is proposed. First- and second-order Riesz transforms [157, 158] at edges positions detected using the Canny edge detector [62] with a dilation operation in the reference and testing images are used. The feature mask is defined as $M = M_1 \oplus M_2$ where M_1 and M_2 are the edge detection applied to the reference and testing images, and $\oplus$ represents the logic OR operation. The RFSIM index is defined as $RFSIM = \prod_{k=1}^{5} D_k$, where

$$D_k = \frac{\sum_{(i,j)\in\Omega} d_k(i,j) \cdot M(i,j)}{\sum_{\forall(i,j)} M(i,j)}$$ (3.29)

and Ω represents the whole image spatial domain. The $d_k(i,j)$ are computed as shown in Eq. (3.30) where the Riesz transforms of the reference and testing images f_k and g_k, respectively:

$$d_k(i, j) = \frac{2f_k(i, j) \cdot g_k(i, j) + c}{f_k^2(i, j) \cdot g_k^2(i, j) + c}, \tag{3.30}$$

where c is a small constant used to avoid numerical instability.

A feature similarity index for image quality assessment (FSIM)

The phase congruency [159], which is a dimensionless measure of the local structure of an image, is used in [134] as a feature to estimate the quality of an image. The phase congruency has been also applied in no-reference sharpness assessment [160].

The phase congruency of the reference and testing images PC_x and PC_y, respectively, are computed as in [159]. The gradient maps G_x and G_y are defined using Scharr operators [154]. The phase congruency and gradient maps of both images are used to compute the following similarity measures:

$$S_{PC}(i, j) = \frac{2PC_x(i, j)PC_y(i, j) + T_1}{PC_x^2(i, j) + PC_y^2(i, j) + T_1} \tag{3.31}$$

$$S_G(i, j) = \frac{2G_x(i, j)G_y(i, j) + T_2}{G_x^2(i, j) + G_y^2(i, j) + T_2}, \tag{3.32}$$

where T_1 and T_2 are positive constants used to increase the stability, and their value depends of the dynamic range of PC and G, respectively. It is known that human visual cortex is sensitive to phase congruent structures, therefore the FSIM is defined as a weighted version of the local similarity $S_L(i, j)$ where the weighting factor is given by $PC_m = \max\left(PC_x(i, j), PC_y(i, j)\right)$. The similarity index FSIM is defined as follows:

$$FSIM = \frac{\sum_{(i,j)\in\Omega} S_L(i, j) \cdot PC_m(i, j)}{\sum_{(i,j)\in\Omega} PC_m(i, j)}, \tag{3.33}$$

where Ω represents the whole image spatial domain. The expression for the local similarity $S_L(i, j)$ is given by

$$S_L(i, j) = [S_{PC}(i, j)]^\alpha \cdot [S_G(i, j)]^\beta. \tag{3.34}$$

The FSIM index on Eq. (3.33) is defined for grayscale images or the luminance component of color images. Since the chrominance component of color images also have an influence on the HVS, a color *FSIM*, namely *FSIM$_C$* is defined using the same procedure described above but this time using both, the luminance and chrominance components of the color image, for this purpose the image is first transformed to the Luminance, In phase, Quadrature (YIQ) color space.

Information Content Weighting for Perceptual Image Quality Assessment

In [131] information weighted MSE, PSNR, and MS-SSIM are defined as follows:

$$IW - MSE = \prod_{k=1}^{M} \left[\frac{\sum_{(i,j)\in\Omega} w^k(i, j)(X^k(i, j) - Y^k(i, j))^2}{\sum_{\forall(i,j)} w^k(i, j)} \right]^{\beta k} \tag{3.35}$$

$$IW - PSNR = 10 \log_{10} \left(\frac{L^2}{IW - MSE} \right) \tag{3.36}$$

$$IW - MS - SSIM = \prod_{k=1}^{M} \left(IW - SSIM^k \right)^{\beta^k}, \tag{3.37}$$

where Ω is the whole image spatial domain, $w(i,j)$ are the information weights defined as in [161], M is the number of scales used within the multi-scale framework, L is the maximum dynamic range, e.g., 255 in an 8 bits/pixel gray-scale image, the β^k are all the same defined in [132], and

$$IW - SSIM^k = \begin{cases} \frac{\sum_{\forall(i,j)} w^k(i,j) c^k(X,Y) s^k(X,Y)}{\sum_{\forall(i,j)} w(i,j)}, & \text{for } k = 1, 2, \ldots, M-1 \\ \frac{1}{N^k} \sum_{(i,j) \in \Omega} l(X,Y) c(X,Y) s(X,Y), & \text{for } k = M, \end{cases} \tag{3.38}$$

where N_k is the number of pixels of the kth scale.

Most apparent distortion (MAD)

In [162] it is assumed that depending on the amount of image distortion, the HVS uses two different approaches in order to extract information from the image. When the distortion is clearly visible, the HVS looks for the image's content information without paying much attention to the distortion; however, when the distortion is small, almost imperceptible, the HVS tries to detect the distortion in the high-quality image. In [162] a method that tries to include these two strategies; an appearance-based and a detection-based is proposed. To determine the perceived distortion in high-quality images, a visibility-weighted local MSE measured in the lightness domain is defined as follows:

$$d_{detect} = \sqrt{\frac{1}{P} \sum_{p} [\xi(p) \cdot D(p)]^2}, \tag{3.39}$$

where $D(p)$ is the MSE computed for every block p of 16×16 pixels, and $\xi(p)$ is the distortion visibility map built using the local contrast map of the reference image and the error image which is defined as the difference between the reference and testing image in the relative lightness domain. Higher values of d_{detect} represent more notorious artifacts in the testing image. For low-quality images, i.e., when the distortion is strong, it is assumed the HVS uses an appearance-based strategy. This is modeled as follows:

$$d_{appear} = \sqrt{\frac{1}{P} \sum_{p} \eta^2(p)}, \tag{3.40}$$

where $\eta(p)$ includes the local differences of the standard deviation, skewness, and kurtosis of twenty subbands (five scales and four orientations) of the reference and testing images. Statistical parameters are defined locally using 16×16 pixel blocks with 75% of overlapping. The statistical comparison is computed as follows:

$$\eta(p) = \sum_{s=1}^{5} \sum_{o=1}^{4} w_s \left[\left| \sigma_{s,o}^{ref}(p) - \sigma_{s,o}^{test}(p) \right| + 2 \left| s_{s,o}^{ref}(p) - s_{s,o}^{test}(p) \right| + \left| \kappa_{s,o}^{ref}(p) - \kappa_{s,o}^{test}(p) \right| \right], \quad (3.41)$$

where $\sigma_{s,o}(p)$, $s_{s,o}(p)$, and $\kappa_{s,o}(p)$ are, respectively, the standard deviation, skewness, and kurtosis of the p block that belongs to the s scale and o orientation subband. The weights w_s are $0.5, 0.75, 1, 5$, and 6 for the finest to coarsest scales, respectively. Finally, the most apparent distortion index (MAD) is defined as the weighted geometric mean of d_{detect} and d_{appear} as follows:

$$MAD = (d_{detect})^{\alpha} \cdot (d_{appear})^{(1-\alpha)}, \quad (3.42)$$

where α is selected based on the level of distortion, e.g., for high-quality images α is close to 1. The optimal technique to define α is still an open problem.

A Visual Saliency-Induced Index (VSI)

In [141] a visual saliency-based image quality index is proposed. The visual saliency (VS) is used as a local feature as well as a weighting factor during pooling the quality index. Several methods have been proposed to estimate the VS of an image, a good compendium of VS methods can be found in [163]. Similarly, several IQA methods involving VS have been proposed [164, 166]. The lack of contrast sensitivity of VS is compensated by including a gradient-based component in the proposed IQA method. To compute the image gradients, first, the RGB image is transformed to the LMN color space, then gradients are computed using Scharr operator [154] on the L image channel. The chrominance channels M, N are used to characterize quality distortions due to color distortions. The visual saliency similarity $S_{VS}(i, j)$, gradient similarity $S_G(i, j)$, and similarity between components $S_C(i, j)$ are defined as follows:

$$S_{VS}(i, j) = \frac{2VS_x(i, j) \cdot VS_y(i, j) + C_1}{VS_x^2(i, j) + VS_y^2(i, j) + C_1} \quad (3.43)$$

$$S_G(i, j) = \frac{2G_x(i, j) \cdot G_y(i, j) + C_2}{G_x^2(i, j) + G_y^2(i, j) + C_2} \quad (3.44)$$

$$S_C(i, j) = \frac{2M_x(i, j) \cdot M_y(i, j) + C_3}{M_x^2(i, j) + M_y^2(i, j) + C_3} \cdot \frac{2N_x(i, j) \cdot N_y(i, j) + C_3}{N_x^2(i, j) + N_y^2(i, j) + C_3}, \quad (3.45)$$

where C_1, C_2, and C_3 are constants to increase computation stability, x and y are the reference and testing images, respectively. Finally, the *VSI* index is defined as follows:

$$VSI = \frac{\sum_{\Omega} S_{VS}(i, j) \cdot [S_G(i, j)]^{\alpha} \cdot [S_C(i, j)]^{\beta} \cdot VS_m(i, j)}{\sum_{\Omega} VS_m(i, j)}, \quad (3.46)$$

where the maximum of the VS between the reference and testing images, i.e., $VS_m = \max\left(VS_x(i, j), VS_y(i, j)\right)$ is used to weight the importance of the three similarity terms. Ω is the whole spatial domain of the testing image, and α, β are used to adjust the relative importance of S_{VS}, S_G, and S_C.

3.5   REDUCED-REFERENCE IMAGE ASSESSMENT

A reduced-reference (RR) quality assessment method based on statistics computed for natural images in the wavelet transform domain is presented in [167]. First, 12 oriented subbands are created with the input image; for this purpose, a steerable pyramid transform with 3-scales and 4-orientations is used. Out of the 12 subbands only 6 are used for extracting features. Subbands are selected such that redundant information is minimized. For each selected subband, the histograms of the coefficients are computed and then a two-parameter generalized Gaussian density (GGD) model is fitted. The two parameters of the GGD and the Kullback–Leibler distance [168] between the wavelet coefficient distributions of the reference and testing images provide 162 bits of information.

An enhancement to the described RR-IQA method is proposed in [169]. In order to reduce the statistical dependencies of multi-scale decomposition coefficients, the divisive normalization transformation (DNT) [170] is applied to the computed wavelet coefficients.

Although wavelets are widely used in IQA methods, they cannot explicitly extract image geometric information; however, the multiscale geometric analysis (MGA) provides a variety of transforms like wavelet, curvelet, bandelet, contourlet, wavelet-based contourlet transforms (WBCT), hybrid wavelets, and directional filter banks (HWD). These transforms extract different features of an image, features that complement to each other providing a better image description. Therefore, in [171] the use of MGA is proposed to extract multichannel structure features. Specifically, in [171] in order to mimic the filter bank-like performance of the HVS, the MGA transforms used to image decomposition and coefficient extraction are curvelet, bandelet, contourlet, WBCT, and HDW. To balance the perceptual importance of coefficients computed with the MGA at different subbands, the modulation transfer function is used to compute the CSF masking coefficients which are used as per-scale weights. Finally, the coefficients with values higher than the Just Noticeable Difference (JND) are used to compute the normalized histogram with as many bins as subbands are used. Once the normalized histograms for the reference and testing image are computed, the image quality metric is computed as follows:

$$Q = \frac{1}{1 + \log_2\left(\frac{S}{Q_o} + 1\right)}, \tag{3.47}$$

where $S = \sum_{n=1}^{L} |P_R(n) - P_T(n)|$, and $P_R(n)$, $P_D(n)$ are the normalized histograms of the reference and testing images, respectively.

CHAPTER 4

No-Reference Image Assessment

There is extensive work in the field of no-reference image quality assessment (NR-IQA). In order to examinate different approaches we have classified these methods as edge-based methods, perceptual-based methods, sparse representation-based methods, among others.

4.1 EDGE-BASED METHODS

Perhaps the most intuitive technique of sharpness estimation is based on the analysis of edges. A smooth edge is obtained when convolving a perfect edge with a blurring kernel, e.g., a Gaussian kernel. This observation has been exploited by several sharpness assessment algorithms either on the spatial or frequency domain. Edge-based methods working on the spatial domain usually use edge contrast and edge extension in their analysis. A smooth edge can also be understood as a perfect edge without its high-frequency components. Therefore, edge-based methods working in the frequency domain usually concentrate their effort on the analysis of the spectral density function around edge pixels.

Methods using kurtosis

In [61], Canny edge detector [62] is used for edge detection, the strength of the selected edges is controlled by the Canny edge detector threshold. For each edge pixel the PDF is obtained as the normalized DCT of the 8×8 block around the edge pixel (excluding the $(0,0)$ term of the spatial frequencies). The 2D kurtosis defined as $Kurt(x) = \frac{m_4(x)}{m_2^2(x)}$ where $m_4(x)$ and $m_2(x)$ are the fourth and second central moments, respectively, is used as local measure of sharpness. The sharpness of the whole image is defined as the average of the kurtoses over all 8×8 blocks.

Method using edge modeling

In [63] an edge sharpness descriptor namely gradient profile sharpness (GPS) is proposed. Given an edge pixel x_o, its 1D gradient profile p is extracted. Although gradient profile is usually modeled using a generalized Gaussian distribution (GGD), in [63] it is modeled with two linear segments, i.e., a triangle. The two linear segments are fitted separately using $P_T(x) = kd_x + b$ where $k \in (k_{left}, k_{right})$ is the slope of the two lines.

The vertical intercept $b = h_T$ where h_T is the maximum value of the gradient profile of x_o. The slopes k_{left} and k_{right} are defined by fitting the edge profile values at each side. Weighted least square is used as fitting method where a Gaussian is used as weighting function. The GPS is defined as the ratio of the height h_t and the edge profile scattering d_T:

$$\eta = \frac{h_T}{d_T} = \frac{h_T}{d_{T_{left}} + d_{T_{right}}} = \frac{h_T}{\left|\frac{h_T}{k_{left}}\right| + \left|\frac{h_T}{k_{right}}\right|} = \frac{\left|k_{left} \cdot k_{right}\right|}{\left|k_{left}\right| + \left|k_{right}\right|}. \tag{4.1}$$

By including the gradient magnitude and the edge spread, the GPS considers the contrast difference of the edge which is an important factor for the HVS.

4.2   METHOD USING IMAGE GRADIENTS

Image gradients are commonly used in sharpness assessment methods, however the relationship between gradients and sharpness perception is not fully understood. In [64] a simple, extremely effective, and fast method for sharpness assessment is proposed. The authors of [64] prove using two experiments that maximum gradients are highly correlated with perceived image sharpness. Therefore, given a grayscale image $I(i, j)$ and its gradients $G(i, j) = \sqrt{(I(i, j) - I(i, j + 1))^2 + (I(i, j) - I(i + 1, j))^2}$ the image sharpness score (ISS) is defined as follows:

$$ISS = MG^\alpha \cdot VG^{1-\alpha}, \tag{4.2}$$

where MG is the maximum gradient of the image, i.e., $MG = \max(G(i, j))$ and VG is the variability of gradients defined as follows:

$$VG = \frac{\max(G(i, j)) - \min(G(i, j))}{\sum_{i,j} \frac{G(i,j)}{W \times H}}, \tag{4.3}$$

where W and H represent the width and height of the image, respectively. The parameter α is experimentally defined as 0.610; however, if we consider that $\min G(i, j)$ is always zero then Eq. (4.2) can be simplified and the parameter α turns out to be irrelevant

$$ISS = \frac{MG}{\sum_{i,j} \frac{G(i,j)}{W \times H}}. \tag{4.4}$$

The proposed method is fast and its results correlate well with images blurred artificially.

Considering an image $I \in \mathbb{R}^{N_1 \times N_2}$, the perceived image I_o can be considered as the linear system $I_o \approx I * h_{HVS}$, where h_{HVS} is the filter simulating the HVS response. The frequency response of h_{HVS} can be approximated by a low-pass filter (LPF) as follows:

$$H_{\omega_c}^n = \begin{cases} (i\omega)^n, & 0 \le \omega \le \omega_c \\ 0, & \text{elsewhere,} \end{cases} \tag{4.5}$$

where ω_c is the cutoff frequency. In [186], the MaxPol package [187] is used to approximated the LPF using a Finite Impulse Response (FIR) expression of eight taps. The horizontal and vertical decomposition of I using the LPF kernel $d_{\omega_c}^n$ are given by

$$\nabla_{\omega_c}^n I = \left[I * d_{\omega_c}^n, I * d_{\omega_c}^n{}^T \right]^T . \tag{4.6}$$

A feature map is calculated in $\ell_{\frac{1}{2}}$-norm using the first and third order derivatives, i.e., $n \in \{1, 3\}$ as follows:

$$M_{\omega_c}^n = \left(\left| \nabla_{\omega_c}^n I(x) \right|^{\frac{1}{2}} + \left| \nabla_{\omega_c}^n I(y) \right|^{\frac{1}{2}} \right)^2 . \tag{4.7}$$

4.3   PERCEPTUAL-BASED METHODS

Methods based on the concept of Just Noticeable Blur (JNB)

The concept of Just Noticeable Difference (JND), first used in the audio processing and coding community [65, 66], has been successfully used in the image processing [67–70]. Two concepts are combined in [71], the JND defined as the minimum amount of a stimulus required to make it perceptible relative to a background, and the Just Noticeable Blur (JNB) defined as the minimum amount of perceivable blurriness around an edge with an intensity contrast higher than the JND. In [71], it is shown that for a given edge contrast C, the probability of detecting blur is given by:

$$P = 1 - e^{-\left| \frac{\omega(e_i)}{\omega_{JNB}(e_i)} \right|^{\beta}}, \tag{4.8}$$

where $\omega(e_i)$ is the width of the edge e_i, $\omega_{JNB}(e_i)$ is the JNB width which depends of the edge contrast computed from the edge profile, and $3.4 \leq \beta \leq 3.8$. In [71] the JNB width is found to be

$$w_{JNB} = \begin{cases} 5, & \text{if } C \leq 50 \\ 3, & \text{otherwise.} \end{cases} \tag{4.9}$$

The probability of detecting blur in a region R having several edges can be written as:

$$P_{blur}(R) = 1 - P_{sharp}(R) = 1 - \prod_{e_i \in R} (1 - P(e_i)). \tag{4.10}$$

A perceptual blur model based on probability summation of [71] has been extended in [28] where the Cumulative Probability of Blur Detection (CPBD) is defined. Given an image I with detected edge positions e_i, the PDF of blur detection PDF_{BD} defined as the normalized histogram of the probabilities found using Eq. (4.8) for every edge position e_i $P_b = P(e_i)$ is defined. Then the CPDF is defined as

$$CPBD = \sum_{P_b=0}^{P_b=P_{JNB}} P_b. \tag{4.11}$$

According to the JNB concept, edge positions with $P_b < P_{JNB}$ cannot be detected; therefore, a higher CPBD score represents a sharper image.

The concept of JNB is also incorporated in [27] to define a perceptual sharpness index (PSI). First, G_x and G_y, the gradients in the horizontal and vertical direction, respectively, are computed using Sobel kernels. Then, edge positions are defined by thresholding the gradient image $G = G_x^2 + G_y^2$, the threshold is defined as $T = \alpha \cdot \overline{G}$, where α is a constant defined experimentally as 4.7, and $\overline{G}$ denotes the mean of G. The width w of edges e that do not deviate more than $\Delta\phi = 8°$ from the horizontal is computed as follows:

$$W(e) = \frac{w_{up}(e) - w_{down}(e)}{\cos(\Delta\phi(e))}, \tag{4.12}$$

where $w_{up}(e)$ and $w_{down}(e)$ are the distances in the intensity profile between the detected pixel e and the maximum I_{max} and minimum I_{min} luminance pixels, respectively. In order to provide a higher relevance to edge pixel with a higher contrast, the JNB is used to compute the new edge width as follows:

$$w_{PSI}(e) = \begin{cases} w(e) - m(e), & \text{if } w(e) \geq w_{JNB} \\ w(e), & \text{otherwise}, \end{cases} \tag{4.13}$$

where $m(e)$ is the edge slope defined as:

$$m(e) = \frac{I_{max}(e) - I_{min}(e)}{w(e)}. \tag{4.14}$$

Since high contrast edges are desired, in Eq. (4.13) edge contrasts above 128 are considered, i.e., $w_{JNB} = 3$. A local sharpness index for image blocks of 32×32 pixels is defined as the reciprocal of the average $w_{PSI}(e)$ of all detected edges in the block. Finally, the global image sharpness is defined as the average of the highest 22nd percentile of the local sharpness indexes of the whole image.

4.4   METHODS BASED ON SPARSE REPRESENTATION

Given a pristine sharp image $\mathbf{I}_o$ and its blurred version $\mathbf{I}$, the blurred image can be represented using an overcomplete dictionary $\mathbf{D}$ [177] as follows:

$$\mathbf{I} = \mathbf{D}x, \qquad s.t. \qquad \|\mathbf{I} - \mathbf{D}x\|_2 \leq \epsilon, \tag{4.15}$$

where x is the representation vector. Additionally, the pure blur signal can be computed as

$$\mathbf{B} = \mathbf{I} - \mathbf{I}_o. \tag{4.16}$$

It we use the sharp image without its mean value, which is a common practice in sparse representation of signals, we have:

$$\mathbf{B} + \overline{\mathbf{I}}_o = \mathbf{I} - \left(\mathbf{I}_o - \overline{\mathbf{I}_o}\right). \tag{4.17}$$

Calling $\mathbf{B}' = \mathbf{B} + \bar{\mathbf{I}}_o$, using the sparse representation of the blur image $\mathbf{I}$, and taking the expectation, we have:

$$E\left[\mathbf{B}'\right] = E\left[\mathbf{I}\right] - E\left[\mathbf{I}_o - \bar{\mathbf{I}}_o\right]. \tag{4.18}$$

Since $E\left[\mathbf{I}_o - \bar{\mathbf{I}}_o\right] = 0$ we have:

$$E[\mathbf{B}'] = E[\mathbf{D}x] \tag{4.19}$$

which shows that the image blur is related with the sparse coefficients x. Since blur is linked to an attenuation of high-frequency content, it is necessary to show the relationship between the energy of an image block and its sparse coefficients. For this purpose, the gradient of an image block b is represented as:

$$b = \mathbf{D}x = \sum_{i=1}^{K} x_i d_i, \tag{4.20}$$

where d_i are the atoms of the overcomplete dictionary $\mathbf{D}$. Taking the ℓ_2 norm to Eq. (4.20) we have:

$$\|b\|_2^2 = \sum_{i=1}^{K} \langle x_i d_i, x_i d_i \rangle = \sum_{i=1}^{K} |x_i|^2 \langle d_i, d_i \rangle. \tag{4.21}$$

Since the atoms of D are orthonormal among themselves, the inner product $\langle d_i, d_i \rangle = 1$; therefore:

$$\|b\|_2^2 = \sum_{i=1}^{K} |x_i|^2. \tag{4.22}$$

Equation (4.22) shows that the energy of a block gradients $\|b\|_2^2$ is defined by the vector of sparse coefficients. If we consider an image composed by M blocks then the sharpness score of the image can be written as follows:

$$P = \frac{1}{M} \sum_{i=1}^{M} \|\mathbf{b}_i\|_2^2. \tag{4.23}$$

In [184] it is shown that P depends of the image content; therefore, a normalization is introduced to obtain a sharpness score independent of the image content

$$Q = \frac{P}{\frac{1}{M} \sum_{i=1}^{M} \sigma_i^2} = \frac{\|\mathbf{b}_i\|_2^2}{\sum_{i=1}^{M} \sigma_i^2}, \tag{4.24}$$

where σ_i^2 is the variance of the ith block of gradients.

4.5   TRANSFORM-BASED METHODS

In [178], a fast image sharpness (FISH) algorithm to estimate the global and local image sharpness is presented. FISH can be understood as a spectral-based method since it uses the concept that perceived sharpness can be estimated by analyzing energy in high-frequency bands. First, the input image is converted to grayscale which is later decomposed into three levels of wavelets subbands by using the biorthogonal wavelets Cohen–Daubechies–Fauraue 9/7. The discrete wavelets subbands at level n, where $n \in [1, 2, 3]$ defined as S_{LH_n}, S_{HL_n}, and S_{HH_n} are used to compute the log-energy of each subband as follows:

$$E_{XY_n} = \log_{10} \left(1 + \frac{1}{N_n} \sum_{I,j} S_{XY_n}^2 (i, j) \right), \tag{4.25}$$

where XY represents the subbands either LH, HL, or HH (the LL subband is not used) and N_n is the number of coefficients per subband at level n. The energy at each decomposition level n is computed as follows:

$$E_n = (1 - \alpha) \frac{E_{LH_n} + E_{HL_n}}{2} + \alpha E_{HH_n}, \tag{4.26}$$

where $\alpha = 0.8$ is used in order to provide a heavier weight to the HH subband. Finally, the global sharpness score of the input image is computed as follows:

$$FISH = \sum_{n=1}^{3} 2^{3-n} E_n. \tag{4.27}$$

In Eq. (4.27) the factor 2^{3-n}, where $n \in [1, 2, 3]$, is designed to provide a heavier weight to finer scales. FISH can also be used to define a local sharpness score which can be used to create a sharpness map of the input image. FISH results correlate well with sharpness perception and its considerable fast due to the use of the discrete wavelet transform.

4.6   METHOD BASED ON LOCAL VARIATIONS

In [179] an input image is first converted to gray scale and then for every image pixel $I_{i,j}$ the maximum local variation (MLV) is computed as follows:

$$\phi \left(I_{y,j} \right) = \max \left| I_{i,j} - I_{x,y} \right|, \tag{4.28}$$

where $x \in [i - 1, i, i + 1]$ and $y \in [j - 1, j, j + 1]$, i.e., $I_{x,y}$ are the eight pixels around the pixel $I_{i,j}$. The authors of [179] show that the MLV provides a better representation of variations than the total variation defined as follows:

$$\gamma \left(I_{I,j} \right) = \sum_{x=i-1}^{i+1} \sum_{y=j-1}^{j+1} \left| I_{i,j} - I_{x,y} \right|. \tag{4.29}$$

Using the MLV for each image pixel a MLV map $\Phi(I)$ is generated. The map $\Phi(I)$ for an image of size $M \times N$ can be express as follows:

$$\Phi(I) = \begin{bmatrix} \phi(I_{1,1}) & \cdots & \phi(I_{1,N}) \\ \vdots & \ddots & \vdots \\ \phi(I_{M,1}) & \cdots & \phi(I_{M,N}) \end{bmatrix}. \tag{4.30}$$

Since HVS is more sensitive to regions with higher variations, a weight defined as the rank of the MLV $\phi(I_{i,j})$ when they are sorted in ascending order is applied. Finally, the sharpness score is defined as the variance of the weighted MLV distribution.

4.7 SINGULAR VALUE DECOMPOSITION-BASED METHODS

SVD has been successfully used in image quality assessment. In [121] a detailed analysis of the singular values of the gradients of an image is presented. Given an $N \times N$ grayscale image block and the corresponding gradients in the horizontal and vertical directions g_x and g_y, respectively; a $[N^2 \times 2]$ matrix $\mathbf{G}$ can be formed using the vectorized versions of $\mathbf{g}_x$, and $\mathbf{g}_y$ as $\mathbf{G} = \left[\mathbf{g}_x(:) \mathbf{g}_y(:) \right]$. The singular values of $\mathbf{G}$ can be found using the compact singular value decomposition as follows:

$$\mathbf{G} = \mathbf{USV}^T = \mathbf{U} \begin{bmatrix} s_1 & 0 \\ 0 & s_2 \end{bmatrix} [\mathbf{v}_1 \ \mathbf{v}_2]^T, \tag{4.31}$$

where s_1 and s_2 are the singular values and $s_1 > s_2 > 0$. s_1 and s_2 represent the energy in the orientations defined by the eigenvectors $\mathbf{v}_1$ and $\mathbf{v}_2$. Through the analysis of several types of image blocks, it is concluded that the singular value s_1 is intimately related with the sharpness of the image block.

Later, noisy image blocks are analyzed. It is shown that the greatest singular value of a noisy image block can be written as follows:

$$\hat{s}_1 \approx \sqrt{s_1^2 + \xi N^2 \sigma^2}, \tag{4.32}$$

where s_1 is the singular value of the noiseless image block, σ^2 is the variance of the noise, and ξ depends of the filter used to compute the image gradients, e.g., $\xi = \frac{3}{16}$ for Sobel kernel.

Based on the previous analysis, in [121] the following sharpness metric is proposed for an image block:

$$H = \frac{s_1}{\epsilon + \sigma^2}, \tag{4.33}$$

where ϵ is a constant experimentally defined as 1. Therefore, the whole image sharpness score is given by

$$H = \frac{1}{M} \sum_{i=1}^{M} H_i, \qquad (4.34)$$

where H_i is the sharpness of each image block. In [121], image block of $N = 16$ pixels are used.

In [180], it is observed that singular value curve (SVC); i.e., the plot of the singular values of a grayscale image I resembles an inverse power characteristic as follows:

$$y = x^{-q}, \qquad (4.35)$$

where $y = S(i)$ are the singular values of I sorted in descending order and i is the index of the singular value $S(i)$, $i \in \{1, 2, \ldots, r\}$ and r is the rank of I. In order to define the value of the exponent q, Eq. (4.35) is written as $M = qN$ where $M = \ln\left(\frac{1}{y}\right)$ and $N = \ln x$; then, the expression for q is defined using least squares to minimize the quadratic error $\epsilon^2 = \sum_{i=1}^{r} (M_i - qN_i)^2$, as follows:

$$q = \frac{\sum_{i=1}^{r} N_i M_i}{\sum_{i=1}^{r} N_i^2} = -\frac{\sum_{i=1}^{r} \ln i \ln S(i)}{\sum_{i=1}^{r} \ln^2 i}, \qquad S(i) > c, \qquad (4.36)$$

where c is experimentally defined, $c = 50$ is used in the experiments reported in [180].

In [181], the input image is first converted to grayscale, then the gradient map is computes as $\mathbf{G} = \frac{|G_x| + |G_y|}{2}$, where $\mathbf{G}_x$ and $\mathbf{G}_y$ are the gradients in the horizontal and vertical directions, respectively. The grayscale image $\mathbf{I}$, the gradient map $\mathbf{G}$, and the saliency map $\mathbf{S}$ are divided into non-overlapping blocks of $m \times n$ pixels; $m = n = 6$ are used in the implementation of [181]. The DC coefficient of the DCT of the gradient blocks $\mathbf{B}_{ij}^{G}$ is set to zero as follows:

$$\mathbf{L}_{ij} = \begin{bmatrix} 0 & D_{1,2} & \cdots & D_{1n} \\ D_{2,1} & D_{m,2} & \cdots & D_{2n} \\ \vdots & \vdots & \ddots & \vdots \\ D_{m,1} & D_{m,2} & \cdots & D_{mn} \end{bmatrix}. \qquad (4.37)$$

The vertical and horizontal differences of the DCT coefficients $\mathbf{L}_{i,j}$ are computed as follows:

$$\mathbf{F}_{i,j}^{H} = \mathbf{L}_{x,y} - \mathbf{L}_{x,y+1} \qquad x \in \{1, \ldots, m\}, \quad y \in \{1, \ldots, n-1\} \qquad (4.38)$$

$$\mathbf{F}_{i,j}^{V} = \mathbf{L}_{x+1,y} - \mathbf{L}_{x,y} \qquad x \in \{1, \ldots, m-1\}, \quad y \in \{1, \ldots, n\}. \qquad (4.39)$$

The vectorized forms of $\mathbf{F}_{i,j}^{H}$ and $\mathbf{F}_{i,j}^{H}$, i.e., $\mathbf{F}_{i,j}^{H}(:)$ and $\mathbf{F}_{i,j}^{V}(:)$ are used to form $\mathbf{F} = \left[\mathbf{F}_{i,j}^{H}(:), \mathbf{F}_{i,j}^{V}(:) \right]$. The singular values of $\mathbf{F}$ are computed by singular value decomposition $\mathbf{F} = \mathbf{U}\mathbf{S}\mathbf{V}^{T}$. It is observed that the singular values s_1, s_2 of a block are smaller for more blurred images; therefore, a response function of singular values (RFSV) inspired in the Harris corner detector [78] is defined as follows:

$$E_{ij} = s_1 \times s_2 - \alpha \, (s_1 + s_2)^2 \,, \tag{4.40}$$

where $\alpha = 0.01$. The sum of RFSV of all the image blocks can be used as sharpness score; however, it is sensible to image content; therefore, the per block image variance σ_{ij}^2 and the per block DCT domain entropy are used to normalized the RFSV and reduce the image content dependency. The per block image variance is defined as follows:

$$\sigma_{ij} = \frac{1}{mn} \sum_{x=1}^{m} \sum_{y=1}^{n} \left(\mathbf{B}_{ij}^I(x, y) - \mu \right)^2 \,, \tag{4.41}$$

where μ represents the mean value of the block $\mathbf{B}_{ij}^I$. The per block DCT-based weight p_{ij} and DCT domain entropy c_{ij} are defined as follows:

$$p_{ij}(x, y) = \frac{L_{ij}(x, y)^2}{\sum_{x=1}^{m} \sum_{y=1}^{n} L_{ij}(x, y)^2} \tag{4.42}$$

$$c_{ij} = - \sum_{x=1}^{m} \sum_{y=1}^{n} p_{ij}(x, y) \log_2 p_{ij}(x, y). \tag{4.43}$$

The SIFT-based saliency map is also divided in $m \times n$ blocks $\mathbf{B}_{ij}^S$, the saliency block weights are defined as follows:

$$W_{ij} = \begin{cases} 1 + e^{\frac{1}{n_{ij}^\beta}}, & n_{ij} \neq 0 \\ 0, & n_{ij} = 0, \end{cases} \tag{4.44}$$

where β is experimentally defined as $\beta = 20$. The final image sharpness score is defined as follows:

$$Score = r \times \frac{\sum_{i=1}^{R} \sum_{j-1}^{K} W_{ij} \cdot E_{ij}}{\sum_{i=1}^{R} \sum_{j=1}^{K} W_{ij} \left(\sigma_{ij}^2 + c_{ij}^2 \right)}, \tag{4.45}$$

where $r = 1$ is used in the implementation of [181] and R, K are, respectively, the number of vertical and horizontal blocks of the input image.

4.8 AUTOREGRESSIVE-BASED METHOD

Based on the hypothesis that image blurring increases the likeness of locally estimated autoregressive (AR) parameters, in [182] a method that exploits the energy- and contrast-differences in the locally estimated AR coefficients is presented. The proposed AR-based Image Sharpness

Metric (ARISM) defines for each image pixel at location (i, j) an 8th order AR model $W_{i,j}$ using the 8-connected neighboring pixels. Using the AR parameters two measures are defined, one that considers the energy as follows:

$$E_{i,j} = |W_{max} - W_{min}|^2 , \tag{4.46}$$

where

$$W_{max} = \max_{(s,t)\in\Omega_{i,j}} (s,t), \tag{4.47}$$

$$W_{min} = \min_{(s,t)\in\Omega_{i,j}} (s,t), \tag{4.48}$$

and

$$\Omega_{i,j} = \{(s,t) \,|\, s \in [i-1, i+1] , t \in [j-1, j+1] , (s,t) \neq (i, j)\} . \tag{4.49}$$

The second measure considers the local contrast as follows:

$$C_{i,j} = \frac{(W_{max} - W_{min})^2}{W_{max}^2 - W_{min}^2} . \tag{4.50}$$

The block-based versions of both measures E and C per image block of $[M \times M]$ pixels are defined as follows:

$$E_{u,v}^{bb} = \frac{1}{M} \sqrt{\sum_{(i,j)\in\Phi_{u,v}} E_{i,j}} \tag{4.51}$$

$$C_{u,v}^{bb} = \frac{1}{M} \sqrt{\sum_{(i,j)\in\Phi_{u,v}} C_{i,j}} , \tag{4.52}$$

where $\Phi_{u,v}$ is the image patch which is defined as follows:

$$\Phi_{u,v} = \{(i, j) \,|\, i \in [(u-1) M + 1, uM] , j \in [(v-1) M + 1, vM]\} , \tag{4.53}$$

where u and v are the indexes of the vertical and horizontal image blocks, respectively.

The largest values of each measure are used to define the sharpness scores ρ_k where $k \in \{E, C, E_{bb}, C_{bb}\}$. The global image sharpness score is defined as a weighted pooling of these scores as follows:

$$\rho = \sum_{k\in\Psi} \Theta_k \rho_k , \tag{4.54}$$

where $\Psi = \{E, C, E_{bb}, C_{bb}\}$ and Θ are weights assigned according to the importance of each measure. The authors in [182] also define a sharpness score that considers the chrominance information of the input image. For this purpose, the image in the YIQ color space is used as follows:

$$\rho_c = \sum_{l\in\{Y,I,Q\}} \Delta l \cdot \rho_l , \tag{4.55}$$

where Δl are weights used to provide relative importance to the YIQ color space components.

4.9   MOMENT-BASED METHOD

In [183] the discrete Tchebichef moments are used to compute the energy associated with image blocks. Given an input image it is converted to grayscale $\mathbf{I}$. The image gradient is computed as $\mathbf{G} = \frac{|\mathbf{G}_x| + |\mathbf{G}_y|}{2}$ where $\mathbf{G}_x$ and $\mathbf{G}_y$ are the gradients in the horizontal and vertical directions, respectively. The grayscale image $\mathbf{I}$ of size $[M \times N]$ and the image gradient $\mathbf{G}$ are divided in non-overlapping blocks of $[D \times D]$ pixels. The number of vertical and horizontal blocks are P and Q, respectively, where $P = \lfloor M/D \rfloor$, $Q = \lfloor N/D \rfloor$, and $\lfloor \cdot \rfloor$ represents the floor operation. The variance σ_{ij}^2 of each image block $\mathbf{B}_{ij}^I$ is computed as well as the Tchebichef moments $\mathbf{T}_{i,j}$ of each gradient block $\mathbf{B}_{ij}^G$. The structure of $\mathbf{T}_{i,j}$ is given by

$$\mathbf{T}_{ij} = \begin{bmatrix} T_{00} & T_{01} & \cdots & T_{0n} \\ T_{10} & T_{11} & \cdots & T_{1n} \\ \vdots & \vdots & \ddots & \vdots \\ T_{m0} & T_{m1} & \cdots & T_{mn,} \end{bmatrix}, \tag{4.56}$$

where $m, n \in \{0, 1, \ldots, (D-1)\}$. The sum of squared non-DC moment (SSM) values are computed as follows:

$$E_{ij} = \sum_{p=0}^{m} \sum_{q=0}^{n} T_{pq}^2 - T_{00}^2. \tag{4.57}$$

Although the block energy E_{ij} decreases monotonically with increasing levels of blurring, i.e., it is a good sharpness measure, it cannot be used directly since it is content dependent. In order to reduce the content dependency on Eq. (4.57), the variance of the image blocks is introduced as follows:

$$S = \frac{\sum_{i=1}^{P} \sum_{j=1}^{Q} E_{ij}}{\sum_{i=1}^{P} \sum_{j=1}^{Q} \sigma_{ij}^2}. \tag{4.58}$$

Finally, in order to include aspects of the HVS, a saliency map is used to weight the contribution of each block; therefore, the blind image blur evaluation score (BIBLE) is defined as follows:

$$S_{BIBLE} = \frac{\sum_{i=1}^{P} \sum_{j=1}^{Q} \widetilde{W}_{ij} \cdot E_{ij}}{\sum_{i=1}^{P} \sum_{j=1}^{Q} \widetilde{W}_{ij} \cdot \sigma_{ij}^2}, \tag{4.59}$$

where $\widetilde{W}_{ij}$ is the resized version of the original saliency map W_{ij}, $\widetilde{W}_{ij}$ has dimensions $[P \times Q]$ which corresponds to the number of blocks in the vertical and horizontal direction.

4.10   MULTIPLE CRITERIA-BASED METHODS

In [41], spectral and spatial properties of the image are used to provide a sharpness score which is called S_3 (Spectral and Spatial Score). Given an input color image $\mathbf{I}$, its grayscale version $\mathbf{X}$ is used to find both measures S_1 and S_2, the former on the frequency domain and the latter in the spatial domain. In order to define S_1, $\mathbf{X}$ is divided in blocks $\mathbf{x}$ of 32×32 pixels with an overlap of 24 pixels between neighboring blocks. First, the luminance-valued block $\mathbf{l}(\mathbf{x}) = (b + k\mathbf{x})^{\gamma}$ is computed, where $b = 0.7656$, $k = 0.0364$, and $\gamma = 2.2$. If the contrast of $\mathbf{x}$ is zero, the measure S_1 is set to zero. The contrast of $\mathbf{x}$ is considered zero if $\max(\mathbf{l}(\mathbf{x})) - \min(\mathbf{l}(\mathbf{x})) \leq T_1$ or $\mu_{\mathbf{l}(\mathbf{x})} \leq T_2$ where $T_1 = 5$, $T_2 = 2$, and $\mu_{\mathbf{l}(\mathbf{x})}$ denotes the mean of $\mathbf{l}(\mathbf{x})$.

When the contrast of $\mathbf{x}$ is greater than zero, S_1 is defined as follows:

$$S_1(\mathbf{x}) = 1 - \frac{1}{1 + e^{\tau_1(\alpha_x - \tau_2)}}, \tag{4.60}$$

where $\tau_1 = -3$, $\tau_2 = 2$, and α_x is computed as the slope of the line $-\alpha \log f + \log \beta$ that best fits the magnitude spectrum of $\mathbf{x}$ as follows:

$$\alpha_x = \min_{\alpha} |\beta f^{-\alpha} - \mathbf{z}_{\mathbf{x}}(f)|_2^2, \tag{4.61}$$

where $\mathbf{z}_{\mathbf{x}}(f)$ is the magnitude spectrum summed across all orientations θ, $\mathbf{z}_{\mathbf{x}}(f)$ is defined as follows:

$$\mathbf{z}_{\mathbf{x}}(f) = \sum_{\theta} |\mathbf{y}_{\mathbf{x}}(f, \theta)|, \tag{4.62}$$

where $\mathbf{y}_{\mathbf{x}}(f, \theta)$ is the 2D Hamming-windowed Discrete Fourier transform of $\mathbf{x}$ expressed in terms of the frequency components f and its orientations θ.

To define the spatial measure S_2 first, the total variation of the block $\mathbf{x}$, i.e., $v(\mathbf{x})$ is defined as follows:

$$v(\mathbf{x}) = \frac{1}{255} \sum_{i,j} |x_i - x_j|, \tag{4.63}$$

where x_i and x_j are eight neighbor pixels in $\mathbf{x}$; therefore, $v(\mathbf{x})$ will present a high value for image blocks with high contrast. The measure S_2 is defined as follows:

$$S_2(\mathbf{x}) = \frac{1}{4} \max_{\xi \in \mathbf{x}} v(\xi), \tag{4.64}$$

where ξ is a 2×2 block of $\mathbf{x}$. The measures S_1 and S_2 for each block $\mathbf{x}$ are assembled to create two maps $S_1(\mathbf{X})$ and $S_2(\mathbf{X})$. A new map S_3 is defined as the weighted geometric mean as follows:

$$S_3(\mathbf{X}) = S_1(\mathbf{X})^{\eta} \times S_2(\mathbf{X})^{1-\eta}, \tag{4.65}$$

where $\eta = 0.5$. Finally, the image sharpness score is defined as the average of the largest 1% values of $S_3(\mathbf{X})$.

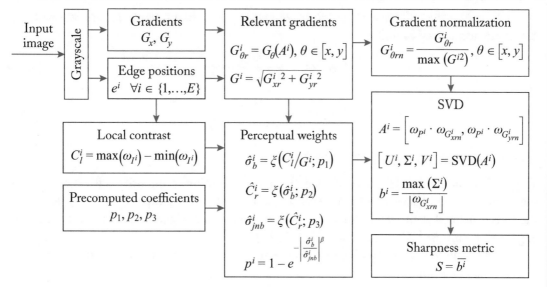

Figure 4.1: Flowchart of the proposed algorithm. Edge positions e^i and gradients $[G_x, G_y]$ are computed using the grayscale version of the input image. The perceptual weights are defined using the pre-computed linear fitting coefficients p_1, p_2, and p_3, the gradients G^i, and the local contrast C_l^i. G^i and C_l^i are defined for the edge positions e^i. ω_{Q^i} represents a block of 5×5 pixel centered in the position $e(i)$ in the image plane Q, $\lfloor \cdot \rfloor$ represents the number of elements of $\cdot$, and β is a constant as in [71].

4.11 SHARPNESS ASSESSMENT BASED ON PERCEPTUALLY WEIGHTED IMAGE GRADIENTS

In the proposed method, sharpness information is extracted using the SVD of perceptually weighted gradients of pixels detected as part of edges in the input image, the weights are defined using the notion of perceptual JNB. The flowchart of the proposed algorithm is shown in Fig. 4.1.

First, the gradients G_x and G_y of the grayscale version of the input image are computed [187]. Since out-of-focus blur degrades the high-frequency content of images, image edges are usually analyzed to extract sharpness information [27, 28, 71, 128, 173]. Pixels detected as edges in the input image are considered as relevant pixels. Edge positions are defined using Canny edge detector [62]. Detected edge positions $e(i)$ are used to extract the sets of relevant gradients in the horizontal and vertical directions $G_{xr}^i = G_x(e(i))$ and $G_{yr}^i = G_y(e(i))$, respectively, where $i \in \{1, \ldots, E\}$ and E is the number of pixels identified as part of the image edges.

To reduce the influence of the image content into the sharpness metric, the relevant gradients are normalized as follows:

$$G_{xrn} = \frac{G_{xr}}{\max(G_{xr}^2 + G_{yr}^2)} \tag{4.66}$$

$$G_{yrn} = \frac{G_{yr}}{\max(G_{xr}^2 + G_{yr}^2)}. \tag{4.67}$$

4.11.1   PERCEPTUAL WEIGHTS

The JND in the field of HV research is defined as the minimum intensity difference between the foreground and background that leads to a noticeable sensory experience [188]. Similarly, the JNB is defined as the minimum amount of blurriness that triggers its perception around an edge with a contrast higher than the JND [16, 71].

According to [71], the probability of detecting a blur distortion has the form of a psychometric function as follows:

$$P = 1 - \exp\left(-\left|\frac{\sigma_b}{\sigma_{jnb}}\right|^\beta\right), \tag{4.68}$$

where σ_b is the standard deviation of the Gaussian blur filter that needs to be applied to an ideal edge to obtain the edge under analysis, σ_{jnb} is the standard deviation of the JNB threshold which is a function of the edge contrast, and β is a constant whose value as reported in [71] is within the range [3.4, 4]. Equation (4.68) usually is computed in terms of edge widths by approximating the real edge width with their horizontal projection [28, 71]. To estimate σ_b and σ_{jnb} for each edge position in the input image we run the following experiment.

A synthetic image having an ideal edge of contrast C_r is blurred with a Gaussian mask with variance σ_b. Edge positions are detected in the blurred image I using Canny edge detector [62]. 5×5 pixels blocks centered at the detected edge pixels are extracted from the blurred image and its gradient magnitude. The blocks with intensity and gradient values, ω_I and ω_G, respectively, are used to compute the local contrast $C_l = \max(\omega_I) - \min(\omega_I)$ and the maximum gradient $G_{max} = \max(\omega_G)$. The mean of C_l and G_{max} are recorded for different values of C_r and σ_b.

Since the quotient C_l/G_{max} is independent of C_r polynomial fitting is used to define the coefficients p_1 of a polynomial of degree 35 for the relation $\sigma_b = \xi(C_l/G_{max}; p_1)$. Under a similar observation, a polynomial of degree 11 is used to fit $C_r/C_l = \xi(\sigma_b; p_2)$. The function ξ represents: $\xi(\sigma_b; p_2) = \sum_{i=0}^{L_{p2}} p_2(i) \cdot \sigma_b^i$ where L_{p_2} is the degree of p_2. Using these two polynomials the real contrast of the edge, C_r, can be estimated using measures of C_l and G_{max} in the edge positions of the input image. The edge contrast C_r is used to compute the perceptual weights defined by Eq. (4.68).

In [71], the standard deviation of the JNB threshold σ_{jnb} for detect blurring in an edge with contrast C_r was established using a subjective test. The coefficients p_3 of a new polynomial of degree 6 are defined such that $\sigma_{jnd} = \xi(C_r; p_3)$; in this case, the values of C_r and σ_{jnb} are those reported in [71]. The degree of the 3 polynomials is defined such that the MSE between the ground-truth curve and the polynomial fit is minimized. Finally, we compute the perceptual weights according to Eq. (4.68) where σ_b and σ_{jnb} are estimated for each edge pixel using local information of the input image.

4.11.2   GLOBAL SHARPNESS INDEX

For each edge pixel position $e(i)$ we define its gradient G^i and its local contrast C_l^i which are used to estimate the standard deviation of the edge $\hat{\sigma}_b$ and the standard deviation of the JNB threshold $\hat{\sigma}_{jnb}$, values that allow us to compute the perceptual weights P^i for each edge pixel according to Eq. (4.68).

For each $e(i)$ we determine its neighboring edge pixels within ω which is a 5×5 pixels block. With the set of edge positions around $e(i)$ we define the column vectors $\omega_{G_{xrn}^i}, \omega_{G_{yrn}^i}$, and ω_{Pi} as the normalized gradients in x and y direction, and the perceptual weights corresponding to the aforementioned edge positions around $e(i)$, respectively.

The SVD has been widely used in image processing and particularly in the estimation of image sharpness [29, 121]. In [121], the authors provide a detailed analysis of the SVD of image gradients to estimate the sharpness level of an image. In [121], the sharpness metric is defined using the greatest singular value of the gradients' matrix. Particularly, for the case of ideal edges, it is straightforward to show that the greatest singular value of the gradient matrix is a function of the edge contrast and the length of the edge. We build on the work of [121] by proposing the formation of the gradient matrix introduced in [121] only with information of the pixels detected as edges in the input image. Additionally, the gradients of the edge pixels are weighted by the probability of perceptual detection of those edge pixels, Eq. (4.68). Finally, the dependence of the greatest singular value on the length of the edge can be easily removed since all the elements in our gradient matrix are part of an edge.

We define the perceptually weighted gradient matrix for the edge pixel $e(i)$ as follows:

$$A_i = \left[\omega_{Pi} \cdot \omega_{G_{xrn}^i}, \omega_{Pi} \cdot \omega_{G_{yrn}^i} \right], \tag{4.69}$$

where $\cdot$ represents the element-by-element product. The sharpness component $b(i)$ associated with the edge pixel position $e(i)$ is defined as follows:

$$b(i) = \frac{\Sigma_{max}}{\left\lfloor \omega_{G_{xrn}^i} \right\rfloor}, \tag{4.70}$$

where $\left\lfloor \omega_{G_{xrn}^i} \right\rfloor$ represents the number of elements of G_{xrn}^i and Σ_{max} is the greatest singular value of the singular value decomposition of A_i.

Finally, after testing different pooling strategies [189], we define the sharpness index of the image as the mean of all the elements of $b(i)$:

$$S = \frac{\sum_i b(i)}{E_r},$$
(4.71)

where E_r represents the number of elements of $b(i)$.

A relevant pixel $e(i)$ with u relevant neighbors within the 5×5 block around it will be present in the computation of $u + 1$ A matrices as defined in Eq. (4.69). Therefore, to avoid the bias in the sharpness index caused by relevant pixels having more neighbors, once a relevant pixel is used, it is void for future computations; this action not only avoids bias in the sharpness index but also speeds-up the process.

4.12   SUMMARY

Although image sharpness is a straightforward task for the human visual system, it is still a challenging problem for computer-based automatic algorithms. Several methods have been developed for assessing blur in images, among which we have frequency-based, perceptual-based, edge analysis-based, phase coherence-based, and sparse representation-based. Here we present a training-free no-reference objective image sharpness assessment method that explicitly exploits the perceptual threshold of the HVS. Perceptual weights are estimated for relevant positions using the local contrast and gradient instead of the traditional edge profiling approach.

CHAPTER 5

Summary and Future Directions

5.1 SUMMARY

Blur is an almost omnipresent effect on image and video. It can represent a challenge in several applications ranging from applications in microscopy imaging to images acquired with telescopes. Image processing and computer vision algorithms usually present unexpected behaviors when the analyzed image is partially or totally blurred. One alternative can be to define a blur map of an image or a video frame, and based on it the image processing or computer vision algorithms can work only over the non-blurred part of the image, or based on the blur map the image can be selectively improved by using blind deconvolution algorithms.

The phenomenon of the main sources of image blurring; i.e, defocus, motion, and atmospheric, were presented. Out-of-focus and motion blur are the most common sources of blurring; therefore, in Chapter 2 a survey of the approaches used to estimate blur maps is presented. Algorithms using gradients, kurtosis, local autocorrelation congruency, transforms, singular value decomposition, the rank of local patches, sparse representation, reblurring, and multi-feature are explained in detail. Additionally, a novel method proposed by the author is presented and compared with state-of-the art algorithms. The dataset of Shi et al. [38], which provides 704 testing images with their corresponding ground truth blurring maps, were used in the quantitative comparison.

Sometimes a single value indicating the degree of blurriness of the whole image is necessary. This type of methods falls within the image quality assessment area. Image quality assessment methods can be categorized as full-reference, reduced-reference, and no-reference methods depending on the amount of information used from the pristine image. Several methods within each of these categories were presented and their pros and cons were made evident. Additionally, a novel no-reference sharpness assessment based on the perceptually-weighted image gradients is presented in detail. The method is compared with eight state-of-the-art no-reference sharpness assessment methods using six publicly available datasets.

In summary, this survey book describes several methods for blur detection as well as image quality assessment. It also provided extensive bibliography. For the reader that wants to examine additional related topics, we provide here more references. Further reading in the utility of machine learning in image enhancement can be found in [197–202]. Surveys on image assessment include [203–208], survey papers on deblurring and image enhancement [209–215], assessments in JPEG-related standards [173, 216–218], deblurring in computational and lens-less cameras [219–222], enhancement for MRI applications in [223–226], and enhancement using graph-based techniques [227–230]. More recently, learning-based methods have been proposed for image quality assessment [231–234].

5.2   FUTURE DIRECTIONS

Once the blur map of an image is defined, it can be used for several image processing tasks. A blur map can be used for image restoration, image classification, image forgery forensics, estimation of the relative depth of objects in the scene, among others.

Blur constitutes a cue of the distance from the camera position to the blurred object; therefore, blur can be exploited to estimate distances. Based on this fact, blur present on images or videos can be part of trending areas of research such as automatic driving vehicles.

As exposed in Chapter 2, most blur detection methods exploit information present on image edges or texture, therefore, detection of blur on image regions that do not have texture, e.g., human skin, still represents a research challenge. Another important aspect that still requires attention is the human perception of blur. Although the psychophysical aspects of human vision have a mature level of understanding, the details about how image blur is understood in an image as a whole are still incomplete.

Sharpness assessment in artificially blurred image has reached acceptable levels of accuracy; however, there is still room for research when considering real blurred images, which usually have localized blurring kernels as well as white Gaussian noise added.

Bibliography

[1] J. Wu, C. Zheng, X. Hu, Y. Wang, and L. Zhang, Realistic rendering of bokeh effect based on optical aberrations, *The Visual Computer*, 26(6–8):555–563, 2010. DOI: 10.1007/s00371-010-0459-5 1

[2] B. A. Barsky, D. R. Horn, S. A. Klein, J. A. Pang, and M. Yu, Camera models and optical systems used in computer graphics: Part I, object-based techniques, *Springer ICCSA*, pages 246–255, 2003. DOI: 10.1007/3-540-44842-x_26 1

[3] G. Mather, Image blur as a pictorial depth cue, *Proc. of the Royal Society of London. Series B: Biological Sciences*, 263(1367):169–172, 1996. DOI: 10.1098/rspb.1996.0027 1

[4] W. G. Chen, N. Nandhakumar, and W. N. Martin, Image motion estimation from motion smear—a new computational model, *IEEE PAMI*, 18(4):412–425, 1996. DOI: 10.1109/34.491622 2

[5] A. C. Aho, K. Donner, S. Helenius, L. O. O. Larsen, and T. Reuter, Visual performance of the toad (bufo bufo) at low light levels: Retinal ganglion cell responses and prey-catching accuracy, *Journal of Comparative Physiology A*, 172(6):671–682, 1993. DOI: 10.1007/bf00195393 2

[6] D. C. Burr and J. Ross, Direct evidence that "speedlines influence motion mechanisms," *Journal of Neuroscience*, 22(19):8661–8664, 2002. DOI: 10.1523/jneurosci.22-19-08661.2002 2

[7] S. Wuerger, R. Shapley, and N. Rubin, On the visually perceived direction of motion by Hans Wallach: 60 years later, *Perception*, 25(11):1317–1367, 1996. DOI: 10.1068/p251317 2

[8] M. C. Roggemann and B. M. Welsh, *Imaging Through Turbulence*, CRC Press, 2018. DOI: 10.1201/9780203751282 3, 4

[9] J. M. Nichols, A. T. Watnik, T. Doster, S. Park, A. Kanaev, L. Cattell, and G. K. Rohde, An optimal transport model for imaging in atmospheric turbulence, *ArXiv:1705.01050*, 2017. 3, 4

[10] D. Li and S. Simske, Atmospheric turbulence degraded-image restoration by kurtosis minimization, *IEEE GRSL*, 6(2):244–247, 2009. DOI: 10.1109/lgrs.2008.2011569 4

[11] S. Li and J. Wang, Adaptive free-space optical communications through turbulence using self-healing bessel beams, *Scientific Reports*, 7:43233, 2017. DOI: 10.1038/srep43233 4

[12] X. Zhu and P. Milanfar, Removing atmospheric turbulence via space-invariant deconvolution, *IEEE PAMI*, 35(1):157–170, 2013. DOI: 10.1109/tpami.2012.82 4

[13] A. H. Mikesell, A. A. Hoag, and J. S. Hall, The scintillation of starlight, *JOSA*, 41(10):689–695, 1951. DOI: 10.1364/josa.41.000689 4

[14] S. Nah, T. H. Kim, and K. M. Lee, Deep multi-scale convolutional neural network for dynamic scene deblurring, *IEEE CVPR*, pages 3883–3891, 2017. DOI: 10.1109/cvpr.2017.35 5

[15] A. C. Bovik and S. T. Acton, Basic linear filtering with application to image enhancement, *The Essential Guide to Image Processing*, pages 225–239, 2009. DOI: 10.1016/b978-0-12-374457-9.00010-x 5

[16] J. Shi, L. Xu, and J. Jia, Just noticeable defocus blur detection and estimation, *IEEE CVPR*, pages 657–665, 2015. DOI: 10.1109/cvpr.2015.7298665 5, 7, 27, 28, 36, 39, 40, 41, 42, 70

[17] S. Golestaneh and L. J. Karam, Spatially-varying blur detection based on multiscale fused and sorted transform coefficients of gradient magnitudes, *IEEE CVPR*, pages 596–605, 2017. DOI: 10.1109/cvpr.2017.71 5, 7, 16, 32, 36, 39, 40, 41, 42

[18] P. Jiang, H. Ling, J. Yu, and J. Peng, Salient region detection by ufo: Uniqueness, focusness and objectness, *IEEE CVPR*, pages 1976–1983, 2013. DOI: 10.1109/iccv.2013.248 5, 7

[19] F. Danieau, A. Guillo, and R. Doré, Attention guidance for immersive video content in head-mounted displays, *IEEE Virtual Reality*, pages 205–206, 2017. DOI: 10.1109/vr.2017.7892248 5

[20] A. Levin, Y. Weiss, F. Durand, and W. Freeman, Efficient marginal likelihood optimization in blind deconvolution, *IEEE CVPR*, pages 2657–2664, 2011. DOI: 10.1109/cvpr.2011.5995308 7

[21] C. T. Shen, W. L. Hwang, and S. C. Pei, Spatially-varying out-of-focus image deblurring with L1-2 optimization and a guided blur map, *IEEE ICASSP*, pages 1069–1072, 2012. DOI: 10.1109/icassp.2012.6288071 7

[22] S. Bae and F. Durand, Defocus magnification. *Computer Graphics Forum*, 26:571–579, Wiley Online Library, 2007. DOI: 10.1111/j.1467-8659.2007.01080.x 7, 31

[23] P. Hsu and B. Yu Chen, Blurred image detection and classification, *Springer International Conference on Multimedia Modeling*, pages 277–286, 2008. DOI: 10.1007/978-3-540-77409-9_26 7

[24] C. Tang, C. Hou, and Z. Song, Depth recovery and refinement from a single image using defocus cues, *Journal of Modern Optics*, 62(6):441–448, 2015. DOI: 10.1080/09500340.2014.967321 7

[25] X. Yi and M. Eramian, LBP-based segmentation of defocus blur, *IEEE TIP*, 25(4):1626–1638, 2016. DOI: 10.1109/tip.2016.2528042 7, 19, 22, 39, 40, 41, 42

[26] K. Bahrami, A. C. Kot, L. Li, and H. Li, Blurred image splicing localization by exposing blur type inconsistency, *IEEE Transactions on IFS*, 10(5):999–1009, 2015. DOI: 10.1109/tifs.2015.2394231 7

[27] C. Feichtenhofer, H. Fassold, and P. Schallauer, A perceptual image sharpness metric based on local edge gradient analysis, *IEEE SPL*, 20(4):379–382, 2013. DOI: 10.1109/lsp.2013.2248711 7, 60, 69

[28] N. D. Narvekar and L. J. Karam, A no-reference image blur metric based on the cumulative probability of blur detection (CPBD), *IEEE TIP*, 20(9):2678–2683, 2011. DOI: 10.1109/tip.2011.2131660 59, 69, 70

[29] X. Zhu and P. Milanfar, Automatic parameter selection for denoising algorithms using a no-reference measure of image content, *IEEE TIP*, 19(12):3116–3132, 2010. DOI: 10.1109/tip.2010.2052820 7, 71

[30] S. Wang, K. Gu, K. Zeng, Z. Wang, and W. Lin, Perceptual screen content image quality assessment and compression, *IEEE ICIP*, pages 1434–1438, 2015. DOI: 10.1109/icip.2015.7351037 7

[31] Q. Yan, Y. Xu, X. Yang, and T. Q. Nguyen, Single image superresolution based on gradient profile sharpness, *IEEE TIP*, 24(10):3187–3202, 2015. DOI: 10.1109/tip.2015.2414877 7

[32] J. F. Cai, H. Ji, C. Liu, and Z. Shen, Blind motion deblurring using multiple images, *Journal of Computational Physics*, 228(14):5057–5071, 2009. DOI: 10.1016/j.jcp.2009.04.022 7

[33] L. Yuan, J. Sun, L. Quan, and H. Y. Shum, Image deblurring with blurred/noisy image pairs, *ACM Transactions on Graphics*, 26:1–10, 2007. DOI: 10.1145/1276377.1276379 7

[34] A. Chakrabarti, T. Zickler, and W. T. Freeman, Analyzing spatially-varying blur, *IEEE CVPR*, pages 2512–2519, 2010. DOI: 10.1109/cvpr.2010.5539954 7, 32, 39, 40, 41, 42

[35] S. Dai and Y. Wu,   Motion from blur.   *IEEE CVPR*, pages 1–8, 2008. DOI: 10.1109/cvpr.2008.4587582 7, 18

[36] R. Liu, Z. Li, and J. Jia, Image partial blur detection and classification, *IEEE CVPR*, pages 1–8, 2008. DOI: 10.1109/cvpr.2008.4587465 7, 12, 13

[37] J. Shi, L. Xu, and J. Jia, Discriminative blur detection features, *IEEE CVPR*, pages 2965–2972, 2014. DOI: 10.1109/cvpr.2014.379 7, 9, 11, 12, 13, 32, 34, 39, 41, 42

[38] J. Shi, Blur detection Dataset, http://www.cse.cuhk.edu.hk/leojia/projects/dblurdetect/index.html, 2014. 8, 9, 14, 17, 18, 20, 22, 23, 28, 30, 31, 32, 34, 35, 36, 37, 38, 39, 40, 41, 73

[39] B. Su, S. Lu, and C. L. Tan, Blurred image region detection and classification, *Proc. of the 19th ACM International Conference on Multimedia*, pages 1397–1400, 2011. DOI: 10.1145/2072298.2072024 7, 18, 39, 40, 41, 42

[40] C. Tang, J. Wu, Y. Hou, P. Wang, and W. Li,   A spectral and spatial approach of coarse-to-fine blurred image region detection,   *IEEE SPL*, 23(11):1652–1656, 2016. DOI: 10.1109/lsp.2016.2611608 7, 30, 31, 32, 36, 39, 40, 41, 42

[41] C. T. Vu, T. D. Phan, and D. M. Chandler,   S3: A spectral and spatial measure of local perceived sharpness in natural images,   *IEEE TIP*, 21(3):934, 2012. DOI: 10.1109/TIP.2011.2169974 7, 13, 15, 67

[42] C. T. Vu, Sharpness maps database, http://vision.eng.shizuoka.ac.jp/s3/#Download, 2012. 15, 42

[43] G. Xu, Y. Quan, and H. Ji, Estimating defocus blur via rank of local patches, *IEEE ICCV*, pages 5371–5379, 2017. DOI: 10.1109/iccv.2017.574 7, 19, 20

[44] X. Zhu, S. Cohen, S. Schiller, and P. Milanfar,   Estimating spatially varying defocus blur from a single image,   *IEEE TIP*, 22(12):4879–4891, 2013. DOI: 10.1109/tip.2013.2279316 7

[45] S. Liu, F. Zhou, and Q. Liao,   Defocus map estimation from a single image based on two-parameter defocus model,   *IEEE TIP*, 25(12):5943–5956, 2016. DOI: 10.1109/tip.2016.2617460 29

[46] J. Thiagarajan, K. Ramamurthy, P. Turaga, and A. Spanias, *Image Understanding Using Sparse Representations*, Morgan & Claypool Publishers, Synthesis Lectures on Image, Video, and Multimedia Processing, 7(1):1–118, 2014. DOI: 10.2200/S00563ED1V01Y201401IVM015 7

[47] M. Hassaballah, A. A. Abdelmgeid, and H. A. Alshazly, Image features detection, description and matching, *Springer Image Feature Detectors and Descriptors*, pages 11–45, 2016. DOI: 10.1007/978-3-319-28854-3_2 7

[48] G. Kulkarni, V. Premraj, V. Ordonez, S. Dhar, S. Li, Y. Choi, A., et al., Babytalk: Understanding and generating simple image descriptions, *IEEE PAMI*, 35(12):2891–2903, 2013. DOI: 10.1109/tpami.2012.162 7

[49] H. Zhao, J. Shi, X. Qi, X. Wang, and J. Jia, Pyramid scene parsing network, *IEEE CVPR*, pages 2881–2890, 2017. DOI: 10.1109/cvpr.2017.660 7

[50] D. Krishnan, T. Tay, and R. Fergus, Blind deconvolution using a normalized sparsity measure, *IEEE CVPR*, pages 233–240, 2011. DOI: 10.1109/cvpr.2011.5995521 7

[51] A. Levin, Y. Weiss, F. Durand, and W. T. Freeman, Understanding and evaluating blind deconvolution algorithms, *IEEE CVPR*, pages 1964–1971, 2009. DOI: 10.1109/cvpr.2009.5206815 7

[52] D. Perrone and P. Favaro, Total variation blind deconvolution: The devil is in the details, *IEEE CVPR*, pages 2909–2916, 2014. DOI: 10.1109/cvpr.2014.372 7, 8

[53] C. Y. Wee and R. Paramesran, Image sharpness measure using eigenvalues, *IEEE CVPR*, pages 840–843, 2008. DOI: 10.1109/icosp.2008.4697259 7

[54] T. Zhu and L. J. Karam, Efficient perceptual-based spatially varying out-of-focus blur, detection, *IEEE ICIP*, pages 2673–2677, 2016. DOI: 10.1109/icip.2016.7532844 7

[55] J. Da Rugna and H. Konik, Automatic blur detection for meta-data extraction in content-based retrieval context, *ISOP Internet Imaging V*, 5304:285–295, 2003. DOI: 10.1117/12.526949 8

[56] A. Levin, R. Fergus, F. Durand, and W. T. Freeman, Deconvolution using natural image priors, *MIT, Computer Science and Artificial Intelligence Laboratory*, 3, 2007. 8

[57] Y. Wang, J. Yang, W. Yin, and Y. Zhang, A new alternating minimization algorithm for total variation image reconstruction, *Journal on Imaging Sciences*, 1(3):248–272, 2008. DOI: 10.1137/080724265 8

[58] D. Krishnan and R. Fergus, Fast image deconvolution using hyper-Laplacian priors, *Advances in Neural Information Processing Systems*, pages 1033–1041, 2009. 8

[59] P. Arbelaez, M. Maire, C. Fowlkes, and J. Malik, Contour detection and hierarchical image segmentation, *IEEE PAMI*, 33(5):898–916, 2010. DOI: 10.1109/tpami.2010.161 9, 10, 11

[60] P. H. Westfall, Kurtosis as peakedness, 1905–2014. R.I.P., *The American Statistician*, 68(3):191–195, 2014. DOI: 10.1080/00031305.2014.917055 9

[61] J. Caviedes and F. Oberti, A new sharpness metric based on local kurtosis, edge and energy information, *Signal Processing: Image Communication*, 19(2):147–161, 2004. DOI: 10.1016/j.image.2003.08.002 9, 57

[62] J. Canny, A computational approach to edge detection, *IEEE PAMI*, (6):679–698, 1986. DOI: 10.1109/tpami.1986.4767851 9, 19, 29, 51, 57, 69, 70

[63] Q. Yan, Y. Xu, and X. Yang, No-reference image blur assessment based on gradient profile sharpness, *IEEE Int. Symp. on Broadband Multimedia Systems and Broadcasting*, pages 1–4, 2013. DOI: 10.1109/bmsb.2013.6621727 57

[64] Y. Zhan and R. Zhang, No-reference image sharpness assessment based on maximum gradient and variability of gradients, *IEEE Transactions on Multimedia*, 20(7):1796–1808, 2017. DOI: 10.1109/tmm.2017.2780770 58

[65] T. Painter and A. Spanias, Perceptual segmentation and component selection for sinusoidal representations of audio, *IEEE Transactions on Speech Audio Processing*, 13(2):149–162, 2005. DOI: 10.1109/tsa.2004.841050 59

[66] A. Spanias, T. Painter, and V. Atti, *Audio Signal Processing and Coding*, Wiley, 2007. 59

[67] L. Karam, N. Sadaka, R. Ferzli, and Z. Ivanovski, An efficient selective perceptual-based super-resolution estimator, *IEEE Transactions on Image Processing*, 20(12):3470–3482, Dec. 2011. 59

[68] N. Sadaka, L. Karam, R. Ferzli, and G. Abousleman, A no-reference perceptual image sharpness metric based on saliency-weighted foveal pooling, In *2008 15th IEEE International Conference on Image Processing*, pages 369–372, 2008. 59

[69] Z. Chen and Hongyi Liu, JND modeling: Approaches and applications, In *2014 19th International Conference on Digital Signal Processing*, pages 827–830, 2014. 59

[70] W. Li, Y. Zhang, X. Yang, A survey of JND models in digital image watermarking, In *2012 Proceedings International Conference on Information Technology and Software Engineering*, pages 765–774, Springer, Berlin, Heidelberg, 2013. 59

[71] R. Ferzli and L. J. Karam, A no-reference objective image sharpness metric based on the notion of just noticeable blur (JNB), *IEEE TIP*, 18(4):717–728, 2009. DOI: 10.1109/tip.2008.2011760 59, 69, 70, 71

[72] C. E. Rasmussen, The infinite Gaussian mixture model, *Advances in Neural Information Processing Systems*, pages 554–560, 2000. 11

[73] A. van der Schaaf and J. H. van Hateren, Modelling the power spectra of natural images: Statistics and Information, *Elsevier Vision Research*, 36(17):2759–2770, 1996. DOI: 10.1016/0042-6989(96)00002-8 12, 32, 33, 34

[74] N. Friedman, D. Geiger, and M. Goldszmidt, Bayesian network classifiers, *Machine Learning*, 29(2–3):131–163, 1997. DOI: 10.1002/9780470400531.eorms0099 12

[75] E. Adelson, E. Simoncelli, and W. T. Freeman, *Pyramids and Multiscale Representations*, Gorea A., Ed., Cambridge University Press, Representations and Vision, pages 3–16, 1991. 12

[76] Q. Yan, L. Xu, J. Shi, and J. Jia, Hierarchical saliency detection, *IEEE CVPR*, pages 1155–1162, 2013. DOI: 10.1109/cvpr.2013.153 12

[77] K. P. Murphy, Y. Weiss, and M. I. Jordan, Loopy belief propagation for approximate inference: An empirical study, *Proc. of the 15th Conference on Uncertainty in AI*, pages 467–475, 1999. 12

[78] C. G. Harris, M. Stephens, et al., A combined corner and edge detector, *Alvey Vision Conference*, 15:10–5244, Citeseer, 1988. DOI: 10.5244/c.2.23 13, 65

[79] D. J. Field and N. Brady, Visual sensitivity, blur and the sources of variability in the amplitude spectra of natural scenes, *Vision Research*, 37(23):3367–3383, 1997. DOI: 10.1016/s0042-6989(97)00181-8 13

[80] D. J. Graham, D. M. Chandler, and D. J. Field, Can the theory of whitening explain the center-surround properties of retinal ganglion cell receptive fields? *Vision Research*, 46(18):2901–2913, 2006. DOI: 10.1016/j.visres.2006.03.008

[81] E. S. L. Gastal and M. M. Oliveira, Domain transform for edge-aware image and video processing, *ACM Transactions on Graphics*, 30(4):69, 2011. DOI: 10.1145/2010324.1964964 18

[82] Levin, D. Lischinski, and Y. Weiss, A closed-form solution to natural image matting, *IEEE PAMI*, 30(2):228–242, 2008. DOI: 10.1109/tpami.2007.1177 19, 28, 29, 30, 31, 36

[83] T. Ojala, M. Pietikäinen, and D. Harwood, A comparative study of texture measures with classification based on featured distributions, *Elsevier, Pattern Recognition*, 29(1):51–59, 1996. DOI: 10.1016/0031-3203(95)00067-4 19

[84] T. Ahonen, A. Hadid, and M. Pietikainen, Face description with local binary patterns: Application to face recognition, *IEEE PAMI*, 12:2037–2041, 2006. DOI: 10.1109/tpami.2006.244 19

[85] T. Ojala, M. Pietikäinen, and T. Mäenpää, Multiresolution gray-scale and rotation in-variant texture classification with local binary patterns, *IEEE PAMI*, 7:971–987, 2002. DOI: 10.1109/tpami.2002.1017623 19, 21

[86] P. Knee, Sparse representations for radar with Matlab® examples, *Synthesis Lectures on Algorithms and Software in Engineering*, Ed. A. Spanias, Morgan & Claypool Publishers, 4(1):1–85, 2012. DOI: 10.2200/s00445ed1v01y201208ase010 22

[87] J. Thiagarajan, K. Ramamurthy, A. Spanias, and D. Frakes, Kernel sparse models for automated tumor segmentation, Patent Issued July 18, 2017, US 9,710,916B2. 22

[88] J. Thiagarajan, K. Ramamurthy, P. Sattigeri, and A. Spanias, Recovering degraded images using ensemble sparse models, US 9,875,428, January 2018. 22

[89] A. M. Bruckstein, D. L. Donoho, and M. Elad, From sparse solutions of systems of equations to sparse modeling of signals and images, *SIAM Review*, 51(1):34–81, 2009. DOI: 10.1137/060657704 23

[90] R. G. Baraniuk, Compressive sensing, *IEEE SPM*, 24(4):118–121, 2007. DOI: 10.1109/MSP.2007.4286571 23

[91] D. L. Donoho and X. Huo, Uncertainty principles and ideal atomic decom-position, *IEEE Transactions on Information Theory*, 47(7):2845–2862, 2001. DOI: 10.1109/18.959265 25

[92] D. L. Donoho and M. Elad, Optimally sparse representation in general (nonorthogo-nal) dictionaries via ℓ^1 minimization, *Proc. of the NAS*, 100(5):2197–2202, 2003. DOI: 10.1073/pnas.0437847100 25

[93] F. Bergeaud and S. Mallat, Matching pursuit of images, *IEEE ICIP*, 1:53–56, 1995. DOI: 10.1109/icip.1995.529037 25

[94] J. J. Thiagarajan, K. N. Ramamurthy, and A. Spanias, Learning stable multilevel dictio-naries for sparse representation of images, *IEEE Transactions on NNLS*, 26:1913–1926, 2013. DOI: 10.1109/tnnls.2014.2361052 25

[95] J. Wang, S. Kwon, and B. Shim, Generalized orthogonal matching pursuit, *IEEE Trans-actions on Signal Processing*, 60(12):6202–6216, 2012. DOI: 10.1109/tsp.2012.2218810 25

[96] K. Ramamurthy, J. Thiagarajan, and A. Spanias, Recovering non-negative and com-bined sparse representations, *Elsevier, Digital Signal Processing*, 26:21–35, 2014. DOI: 10.1016/j.dsp.2013.11.003 25

[97] D. L. Donoho, I. Drori, Y. Tsaig, and J. L. Starck, Sparse solution of underdetermined systems of linear equations by stagewise orthogonal matching pursuit, *IEEE Transactions on Information Theory*, 58(2):1094–1121, 2012. DOI: 10.1109/TIT.2011.2173241 26

[98] D. Needell and R. Vershynin, Uniform uncertainty principle and signal recovery via regularized orthogonal matching pursuit, *Foundations of Computational Mathematics*, 9(3):317–334, 2009. DOI: 10.1007/s10208-008-9031-3 26

[99] D. Needell and J. A. Tropp, CoSaMP: Iterative signal recovery from incomplete and inaccurate samples, *Applied and Computational Harmonic Analysis*, 26(3):301–321, 2009. DOI: 10.1016/j.acha.2008.07.002 26

[100] E. J. Candès and B. Recht, Exact matrix completion via convex optimization, *Foundations of Computational Mathematics*, 9(6):717, 2009. DOI: 10.1007/s10208-009-9045-5 26

[101] E. J. Candes, The restricted isometry property and its implications for compressed sensing, *Comptes Rendus Mathematique*, 346(9–10):589–592, 2008. DOI: 10.1016/j.crma.2008.03.014 27

[102] D. L. Donoho et al., Compressed sensing, *IEEE Transactions on Information Theory*, 52(4):1289–1306, 2006. DOI: 10.1109/tit.2006.871582 27

[103] R. Tibshirani, Regression shrinkage and selection via the lasso, *Journal of the Royal Statistical Society: Series B (Methodological)*, 58(1):267–288, 1996. DOI: 10.1111/j.2517-6161.1996.tb02080.x 27

[104] M. A. T. Figueiredo and R. D. Nowak, An EM algorithm for wavelet-based image restoration, *IEEE TIP*, 12(8):906–916, 2003. DOI: 10.1109/tip.2003.814255 27

[105] M. Grant and S. Boyd, CVX: Matlab software for disciplined convex programming, http://cvxr.com/cvx, version 2.1, March, 2014. 27

[106] E. Candes and J. Romberg, l1-magic: Recovery of sparse signals via convex programming, www.acm.caltech.edu/l1magic/downloads/l1magic.pdf, pages 4–14, 2005. 27

[107] J. Thiagarajan, K. Ramamurthy, and A. Spanias, Learning stable multilevel dictionaries for sparse representations, *IEEE Transactions on Neural Network and Learning Systems*, 26(9):1913–1926, 2014. DOI: 10.1109/tnnls.2014.2361052 27

[108] J. Thiagarajan and A. Spanias, Learning dictionaries for local sparse coding in image classification, *IEEE Asilomar Conference on Signals, Systems, and Computers*, pages 2014–2018, 2011. DOI: 10.1109/acssc.2011.6190379 27

[109] D. J. Chen, H. T. Chen, and L. W. Chang, Fast defocus map estimation, *IEEE ICIP*, pages 3962–3966, 2016. DOI: 10.1109/icip.2016.7533103 29

[110] R. Achanta, A. Shaji, K. Smith, A. Lucchi, P. Fua, and S. Süsstrunk, SLIC superpixels compared to state-of-the-art superpixel methods, *IEEE PAMI*, 34(11):2274–2282, 2012. DOI: 10.1109/tpami.2012.120 29, 30

[111] D. Zhou, O. Bousquet, T. N. Lal, J. Weston, and B. Schölkopf, Learning with local and global consistency, *Advances in NIPS*, pages 321–328, 2004. 29

[112] A. Karaali and C. R. Jung, Edge-based defocus blur estimation with adaptive scale selection, *IEEE TIP*, 27(3):1126–1137, 2017. DOI: 10.1109/tip.2017.2771563 29

[113] C. Tang, C. Hou, and Z. Song, Defocus map estimation from a single image via spectrum contrast, *OSA Optics Letters*, 38(10):1706–1708, 2013. DOI: 10.1364/ol.38.001706 31, 32, 33, 38

[114] S. Zhuo and T. Sim, Defocus map estimation from a single image, *Elsevier Pattern Recognition*, 44(9):1852–1858, 2011. DOI: 10.1016/j.patcog.2011.03.009 7, 28, 29, 30, 31, 36, 39, 40, 41, 42

[115] J. Andrade, P. Turaga, and A. Spanias, Spatially-varying sharpness map estimation based on the quotient of spectral bands, *IEEE ICIP*, pages 4020–4024, 2019. DOI: 10.1109/icip.2019.8803406 33

[116] D. Martin, C. Fowlkes, D. Tal, and J. Malik, A database of human segmented natural images and its application to evaluating segmentation algorithms and measuring ecological statistics, *IEEE ICCV*, 2:416–423, 2001. DOI: 10.1109/iccv.2001.937655 33, 34

[117] C. T. Vu, T. D. Phan, and D. M. Chandler, S3: A spectral and spatial measure of local perceived sharpness in natural images, *IEEE TIP*, 21(3):934–945, 2012. DOI: 10.1109/TIP.2011.2169974 32, 39, 40, 41, 42

[118] E. P. Simoncelli and B. A. Olshausen, Natural image statistics and neural representation, *Annual Review of Neuroscience*, 24(1):1193–1216, 2001. DOI: 10.1146/annurev.neuro.24.1.1193 33

[119] K. He, J. Sun, and X. Tang, Guided image filtering, *IEEE PAMI*, 35(6):1397–1409, 2013. DOI: 10.1109/tpami.2012.213 35

[120] X. Liu, M. Tanaka, and M. Okutomi, Noise level estimation using weak textured patches of a single noisy image, *IEEE ICIP*, pages 665–668, 2012. DOI: 10.1109/icip.2012.6466947 38

[121] X. Zhu and P. Milanfar, A no-reference sharpness metric sensitive to blur and noise, *International Workshop on Quality of Multimedia Experience*, pages 64–69, 2009. DOI: 10.1109/qomex.2009.5246976 38, 63, 64, 71

[122] J. Davis and M. Goadrich, The relationship between precision-recall and ROC curves, *International Conference on Machine Learning*, pages 233–240, 2006. DOI: 10.1145/1143844.1143874 40

[123] Z. Wang, A. C. Bovik, H. R. Sheikh, and E. P. Simoncelli, Image quality assessment: From error visibility to structural similarity, *IEEE TIP*, 13(4):600–612, 2004. DOI: 10.1109/tip.2003.819861 40, 45

[124] Z. Wang and A. C. Bovik, *Modern Image Quality Assessment,* Morgan & Claypool, San Rafael, CA, 2006. DOI: 10.2200/S00010ED1V01Y200508IVM003 45

[125] U. Scherhag, C. Rathgeb, J. Merkle, R. Breithaupt, and C. Busch, Face recognition systems under morphing attacks: A survey, *IEEE Access*, 7:23012–23026, 2019. DOI: 10.1109/access.2019.2899367 45

[126] Y. Niu, L. Lin, Y. Chen, and L. Ke, Machine learning-based framework for saliency detection in distorted images, *Multimedia Tools and Applications*, 76(24):26329–26353, 2017. DOI: 10.1007/s11042-016-4128-1 45

[127] B. Ma, L. Huang, J. Shen, L. Shao, M. H. Yang, and F. Porikli, Visual tracking under motion blur, *IEEE TIP*, 25(12):5867–5876, 2016. DOI: 10.1109/tip.2016.2615812 45

[128] J. Guan, W. Zhang, J. Gu, and H. Ren, No-reference blur assessment based on edge modeling, *Elsevier Journal of Visual Communication and Image Representation*, 29:1–7, 2015. DOI: 10.1016/j.jvcir.2015.01.007 45, 69

[129] A. Liu, W. Lin, and M. Narwaria, Image quality assessment based on gradient similarity, *IEEE TIP*, 21(4):1500–1512, 2011. DOI: 10.1109/tip.2011.2175935 48, 49

[130] H. R. Sheikh, A. C. Bovik, and G. De Veciana, An information fidelity criterion for image quality assessment using natural scene statistics, *IEEE TIP*, 14(12):2117–2128, 2005. DOI: 10.1109/tip.2005.859389 45, 46

[131] Z. Wang and Q. Li, Information content weighting for perceptual image quality assessment, *IEEE TIP*, 20(5):1185–1198, 2011. DOI: 10.1109/tip.2010.2092435 45, 49, 52

[132] Z. Wang, E. P. Simoncelli, and A. C. Bovik, Multiscale structural similarity for image quality assessment, *Asilomar Conference on Signals, Systems, and Computers*, 2(1398–1402), 2003. DOI: 10.1109/acssc.2003.1292216 45, 49, 51, 53

[133] W. Xue, L. Zhang, X. Mou, and A. C. Bovik, Gradient magnitude similarity deviation: A highly efficient perceptual image quality index, *IEEE TIP*, 23(2):684–695, 2013. DOI: 10.1109/tip.2013.2293423 49

[134] L. Zhang, L. Zhang, X. Mou, D. Zhang, et al., FSIM: A feature similarity index for image quality assessment, *IEEE TIP*, 20(8):2378–2386, 2011. DOI: 10.1109/tip.2011.2109730 45, 52

[135] Z. Wang and A. C. Bovik, Mean squared error: Love it or leave it? a new look at signal fidelity measures, *IEEE Signal Processing Magazine*, 26(1):98–117, 2009. DOI: 10.1109/msp.2008.930649 45

[136] Z. Wang and A. C. Bovik, A universal image quality index, *IEEE Signal Processing Letters*, 9(3):81–84, 2002. DOI: 10.1109/97.995823 45, 48, 49, 51

[137] H. R. Sheikh and A. C. Bovik. Image information and visual quality, *IEEE TIP*, 15(2):430–444, 2006. DOI: 10.1109/tip.2005.859378 45

[138] L. Zhang, L. Zhang, and X. Mou, RFSIM: A feature based image quality assessment metric using Riesz transforms, *IEEE ICIP*, pages 321–324, 2010. DOI: 10.1109/icip.2010.5649275 46, 51

[139] L. Zhang and H. Li, SR-SIM: A fast and high performance IQA index based on spectral residual, *IEEE ICIP*, pages 1473–1476, 2012. DOI: 10.1109/icip.2012.6467149 46, 49, 50

[140] M. J. Chen and A. C. Bovik, No-reference image blur assessment using multiscale gradient, *EURASIP JIVP*, 2011(1):3, 2011. DOI: 10.1186/1687-5281-2011-3 46

[141] L. Zhang, Y. Shen, and H. Li, VSI: A visual saliency-induced index for perceptual image quality assessment, *IEEE TIP*, 23(10):4270–4281, 2014. DOI: 10.1109/tip.2014.2346028 46, 54

[142] A. M. Eskicioglu and P. S. Fisher, Image quality measures and their performance, *IEEE Transactions on Communications*, 43(12):2959–2965, 1995. DOI: 10.1109/26.477498 46

[143] T. Mitsa and K. L. Varkur, Evaluation of contrast sensitivity functions for the formulation of quality measures incorporated in halftoning algorithms, *IEEE ICASSP*, 5:301–304, 1993. DOI: 10.1109/icassp.1993.319807 47

[144] J. O. Limb, Distortion criteria of the human viewer, *IEEE Transactions on Systems, Man, and Cybernetics*, 9(12):778–793, 1979. DOI: 10.1109/tsmc.1979.4310129 47

[145] A. Samet, M. Ayed, N. Masmoudi, and L. Khriji, New perceptual image quality assessment metric, *Asian Journal of Information Technology*, 4(11):996–1000, 2005. 47

[146] J. C. Brailean, B. J. Sullivan, C. T. Chen, and M. L. Giger, Evaluating the EM algorithm for image processing using a human visual fidelity criterion, *IEEE ICASSP*, pages 2957–2960, 1991. DOI: 10.1109/icassp.1991.151023 47

[147] A. Beghdadi and R. Iordache, Image quality assessment using the joint spatial/spatial-frequency representation, *EURASIP JASP*, page 40, 2006. 47
DOI: 10.1155/asp/2006/80537

[148] K. Egiazarian, J. Astola, N. Ponomarenko, V. Lukin, F. Battisti, and M. Carli, New full-reference quality metrics based on HVS, *Proc. of the 2nd International Workshop on Video Processing and Quality Metrics*, 4, 2006. 47

[149] A. Liu, W. Lin, and M. Narwaria, Image quality assessment based on gradient similarity, *IEEE TIP*, 21(4):1500–1512, 2012. DOI: 10.1109/tip.2011.2175935 45, 46

[150] W. Xue, L. Zhang, X. Mou, and A. C. Bovik, Gradient magnitude similarity deviation: A highly efficient perceptual image quality index, *IEEE TIP*, 23(2):684–695, 2014. DOI: 10.1109/tip.2013.2293423 45, 46

[151] J. M. S. Prewitt, Object enhancement and extraction, *Picture Processing and Psychopictorics*, 10(1):15–19, 1970. 49

[152] Z. Wang, Rate scalable foveated image and video communications, Ph.D. thesis, Department of ECE, The University of Texas at Austin, December 2001. 49, 50

[153] Z. Wang, A. C. Bovik, and L. Lu, Why is image quality assessment so difficult?, *IEEE ICASSP*, 4:3313–3316, 2002. DOI: 10.1109/ICASSP.2002.5745362 49, 50, 51

[154] H. Scharr, Optimal operators in digital image processing, Ph.D. thesis, Ruprecht-Karls-Universität Heidelberg, Heidelberg, Germany, 2000. 50, 52, 54

[155] D. Marr, *Vision: A Computational Investigation Into the Human Representation and Processing of Visual Information*, MIT Press, 2010. DOI: 10.7551/mitpress/9780262514620.001.0001 51

[156] P. Maragos, *Morphological Filtering for Image Enhancement and Feature Detection*, Elsevier, The Image and Video Processing Handbook, pages 135–156, 2005. DOI: 10.1016/B978-012119792-6/50072-3 51

[157] M. Felsberg and G. Sommer, The monogenic signal, *IEEE Transactions on Signal Processing*, 49(12):3136–3144, 2001. DOI: 10.1109/78.969520 51

[158] L. Wietzke, O. Fleischmann, and G. Sommer, 2D signal analysis by generalized Hilbert transforms, *Computer Vision ECCV*, pages 638–649, 2008. DOI: 10.1007/978-3-540-88688-4_47 51

[159] P. Kovesi et al., Image features from phase congruency, *Videre: Journal of Computer Vision Research*, 1(3):1–26, 1999. 52

[160] Z. Wang and E. P. Simoncelli, Local phase coherence and the perception of blur, *Advances in NIPS*, pages 1435–1442, 2004. 52

[161] H. R. Sheikh and A. C. Bovik, Image information and visual quality *IEEE TIP*, 15(2):430–444, 2006. DOI: 10.1109/tip.2005.859378 53

[162] E. C. Larson and D. M. Chandler, Most apparent distortion: Full-reference image quality assessment and the role of strategy, *Journal of Electronic Imaging*, 19(1):1–6, 2010. DOI: 10.1117/1.3267105 53

[163] A. Borji and L. Itti, State-of-the-art in visual attention modeling, *IEEE PAMI*, 35(1):185–207, 2012. DOI: 10.1109/tpami.2012.89 54

[164] I. Gkioulekas, G. Evangelopoulos, and P. Maragos, Spatial Bayesian surprise for image saliency and quality assessment, *IEEE ICIP*, pages 1081–1084, 2010. DOI: 10.1109/icip.2010.5650991 54

[165] A. K. Moorthy and A. C. Bovik, Visual importance pooling for image quality assessment, *IEEE JSTSP*, 3(2):193–201, 2009. DOI: 10.1109/jstsp.2009.2015374

[166] Y. Tong, H. Konik, F. Cheikh, and A. Tremeau, Full reference image quality assessment based on saliency map analysis, *Journal of Imaging Science and Technology*, 54(3):30503:1–30503:14, 2010. DOI: 10.2352/j.imagingsci.technol.2010.54.3.030503 54

[167] Z. Wang, G. Wu, H. R. Sheikh, E. P. Simoncelli, E. H. Yang, and A. C. Bovik, Quality-aware images, *IEEE TIP*, 15(6):1680–1689, 2006. DOI: 10.1109/tip.2005.864165 55

[168] T. M. Cover and J. A. Thomas, *Elements of Information Theory*, John Wiley & Sons, 2012. 55

[169] Q. Li and Z. Wang, Reduced-reference image quality assessment using divisive normalization-based image representation, *IEEE JSTSP*, 3(2):202–211, 2009. DOI: 10.1109/jstsp.2009.2014497 55

[170] O. Schwartz and E. P. Simoncelli, Natural signal statistics and sensory gain control, *Nature Neuroscience*, 4(8):819–825, 2001. DOI: 10.1038/90526 55

[171] X. Gao, W. Lu, D. Tao, and X. Li, Image quality assessment based on multiscale geometric analysis, *IEEE TIP*, 18(7):1409–1423, 2009. DOI: 10.1109/tip.2009.2018014 55

[172] M. L. T. Chae Postek and A. E. Vladár, Image sharpness measurement in scanning electron microscopy part I, *Scanning: The Journal of Scanning Microscopies*, 20(1):1–9, 1998. DOI: 10.1002/sca.1998.4950200101

[173] P. Marziliano, F. Dufaux, S. Winkler, and T. Ebrahimi, Perceptual blur and ringing metrics: Application to JPEG2000, *Signal Processing: Image Communication*, 19(2):163–172, 2004. DOI: 10.1016/j.image.2003.08.003 69, 74

[174] G. Blanchet, L. Moisan, and B. Rougé, Measuring the global phase coherence of an image, *IEEE ICIP*, pages 1176–1179, 2008. DOI: 10.1109/icip.2008.4711970

[175] G. Blanchet and L. Moisan, An explicit sharpness index related to global phase coherence, *IEEE ICASSP*, pages 1065–1068, 2012. DOI: 10.1109/icassp.2012.6288070

[176] R. Hassen, Z. Wang, and M. M. A. Salama, Image sharpness assessment based on local phase coherence, *IEEE TIP*, 22(7):2798–2810, 2013. DOI: 10.1109/tip.2013.2251643

[177] K. Ramamurthy, J. Thiagarajan, A. Spanias, and P. Sattigeri, Boosted dictionaries for image restoration based on sparse representations, *IEEE ICASSP*, pages 1583–1587, 2013. DOI: 10.1109/icassp.2013.6637918 60

[178] P. V. Vu and D. M. Chandler, A fast wavelet-based algorithm for global and local image sharpness estimation, *IEEE Signal Processing Letters*, pages 423–426, 2012. DOI: 10.1109/lsp.2012.2199980 62

[179] K. Bahrami and A. C. Kot, A fast approach for no-reference image sharpness assessment based on maximum local variation, *IEEE Signal Processing Letters*, 21(6):751–755, 2014. DOI: 10.1109/lsp.2014.2314487 62

[180] Q. Sang, H. Qi, X. Wu, C. Li, and A. C. Bovik, No-reference image blur index based on singular value curve, *Journal of VCIR*, 25(7):1625–1630, 2014. DOI: 10.1016/j.jvcir.2014.08.002 64

[181] S. Zhang, P. Li, X. Xu, L. Li, and C. Chang, No-reference image blur assessment based on response function of singular values, *Multidisciplinary Digital Publishing Institute, Symmetry*, 10(8):304, 2018. DOI: 10.3390/sym10080304 64, 65

[182] K. Gu, G. Zhai, W. Lin, X. Yang, and W. Zhang, No-reference image sharpness assessment in autoregressive parameter space, *IEEE TIP*, 24(10):3218–3231, 2015. DOI: 10.1109/tip.2015.2439035 65, 66

[183] L. Li, W. Lin, X. Wang, G. Yang, K. Bahramiand, and A. C. Kot, No-reference image blur assessment based on discrete orthogonal moments, *IEEE Transactions on Cybernetics*, 46(1):39–50, 2015. DOI: 10.1109/tcyb.2015.2392129 67

[184] L. Li, D. Wu, J. Wu, H. Li, W. Lin, and A. C. Kot, Image sharpness assessment by sparse representation, *IEEE Transactions on Multimedia*, 18(6):1085–1097, 2016. DOI: 10.1109/tmm.2016.2545398 61

[185] G. Gvozden, S. Grgic, and M. Grgic, Blind image sharpness assessment based on local contrast map statistics, *Journal of VCIR*, 50:145–158, 2018. DOI: 10.1016/j.jvcir.2017.11.017

[186] M. S. Hosseini and K. N. Plataniotis, Image sharpness metric based on MaxPol convolution kernels, *IEEE ICIP*, pages 296–300, 2018. DOI: 10.1109/icip.2018.8451488 59

[187] M. S. Hosseini and K. N. Plataniotis, Finite differences in forward and inverse imaging problems: MaxPol design, *SIAM Journal on Imaging Sciences*, 10(4):1963–1996, 2017. DOI: 10.1137/17m1118452 59, 69

[188] T. N. Pappas, R. J. Safranek, and J. Chen, Perceptual criteria for image quality evaluation, *Handbook of Image and Video Processing*, pages 669–684, New York, Academic, 2000. DOI: 10.1016/b978-012119792-6/50118-2 70

[189] Z. Wang and X. Shang, Spatial pooling strategies for perceptual image quality assessment, *IEEE ICIP*, pages 2945–2948, 2006. DOI: 10.1109/icip.2006.313136 72

[190] R. Franzen, Kodak lossless true color image suite PhotoCD PCD0992. http://r0k.us/graphics/kodak

[191] N. Ponomarenko, O. Ieremeiev, V. Lukin, K. Egiazarian, L. Jin, J. Astola, B. Vozel, K. Chehdi, M. Carli, F. Battisti, et al., Color image database TID2013: Peculiarities and preliminary results, *European Workshop on Visual Information Processing*, pages 106–111, 2013.

[192] N. Ponomarenko, V. Lukin, A. Zelensky, K. Egiazarian, M. Carli, and F. Battisti, TID2008—a database for evaluation of full-reference visual quality assessment metrics, *Advances of Modern Radioelectronics*, 10(4):30–45, 2009.

[193] P. Le Callet and F. Autrusseau, Subjective quality assessment IRCCyN/IVC database, http://www.irccyn.ec-nantes.fr/ivcdb/

[194] D. M. Chandler and S. S. Hemami, VSNR: A wavelet-based visual signal-to-noise ratio for natural images, *IEEE TIP*, 16(9):2284–2298, 2007. DOI: 10.1109/tip.2007.901820

[195] H. R. Sheikh, M. F. Sabir, and A. C. Bovik, A statistical evaluation of recent full reference image quality assessment algorithms, *IEEE TIP*, 15(11):3440–3451, 2006. DOI: 10.1109/tip.2006.881959

[196] J. Antkowiak, T. D. F. Jamal Baina, F. V. Baroncini, N. Chateau, F. FranceTelecom, A. C. F. Pessoa, et al., Final report from the video quality experts group on the validation of objective models of video quality assessment, March, 2000.

[197] Y-T. Zhou, R. Chellappa, A. Vaid, and B. K. Jenkins, Image restoration using a neural network, *IEEE Transactions on Acoustics Speech Signal Processing*, 36(7):1141–1151, 1988. DOI: 10.1109/29.1641 73

[198] S. Dodge and L. J. Karam, Understanding how image quality affects deep neural networks, *IEEE 8th International Conference on Quality of Multimedia Experience (QoMEX)*, pages 1–6, 2016. DOI: 10.1109/qomex.2016.7498955 73

[199] C. Tian, L. Fei, W. Zheng, Y. Xu, W. Zuo, and C-W. Lin, Deep learning on image denoising: An overview, *Elsevier Neural Networks*, 131:251–275, 2020. DOI: 10.1016/j.neunet.2020.07.025 73

[200] A. Chakrabarti, A neural approach to blind motion deblurring, *Computer Vision ECCV*, pages 221–235, 2016. DOI: 10.1007/978-3-319-46487-9_14 73

[201] T. M. Nimisha, S. A. Kumar, and A. N. Rajagopalan, Blur-invariant deep learning for blind-deblurring, *IEEE International Conference on Computer Vision*, pages 4752–4760, 2017. DOI: 10.1109/iccv.2017.509 73

[202] J. Koh, J. Lee, Jangho, and S. Yoon, Single-image deblurring with neural networks: A comparative survey, *Elsevier Computer Vision and Image Understanding*, 203:103134, 2021. DOI: 10.1016/j.cviu.2020.103134 73

[203] Y. Deng, C. C. Loy, and X. Tang, Image aesthetic assessment: An experimental survey, *IEEE Signal Processing Magazine*, 34(4):80–106, 2017. DOI: 10.1109/msp.2017.2696576 73

[204] A. G. George and A. K. Prabavathy, A survey on different approaches used in image quality assessment, *International Journal of Emerging Technology and Advanced Engineering*, 3(2):197–203, 2014. 73

[205] L. He, F. Gao, W. Hou, and L. Hao, Objective image quality assessment: A survey, *International Journal of Computer Mathematics*, 91(11):2374–2388, 2014. DOI: 10.1080/00207160.2013.816415 73

[206] B. Hua, L. Li, J. Wu, and J. Qian, Subjective and objective quality assessment for image restoration: A critical survey, *Elsevier Signal Processing: Image Communication*, 85:115839, 2020. DOI: 10.1016/j.image.2020.115839 73

[207] L. J. Karam, T. Ebrahimi, S. Hemami, T. Pappas, R. Safranek, Z. Wang, and A. B. Watson, Introduction to the issue on visual media quality assessment, *IEEE Journal on Special Topics in Signal Processing*, Special Issue on Visual Media Quality Assessment, 3(2):189–192, April 2009. DOI: 10.1109/JSTSP.2009.2015485 73

[208] T. Ebrahimi, L. Karam, F. Pereira, K. El-Maleh, and I. Burnett, The quality of multimedia: challenges and trends, *IEEE Signal Processing Magazine*, pages 17 and 148, Nov. 2011. DOI: 10.1109/MSP.2011.942546 73

[209] F. Vankawala, A. Ganatra, and A. Patel, A survey on different image deblurring techniques, *International Journal of Computer Applications*, 116(13):15–18, 2015. DOI: 10.5120/20396-2697 73

[210] M. Poulose, et al., Literature survey on image deblurring techniques, *International Journal of Computer Applications Technology and Research*, 2(3):286–288, 2013. DOI: 10.7753/ijcatr0203.1014 73

[211] A. N. Rajagopalan and R. Chellappa, *Motion Deblurring: Algorithms and Systems*, Cambridge University Press, 2014. DOI: 10.1017/CBO9781107360181 73

[212] Y. R. Sankaraiah and A. Varadarajan, Deblurring techniques—a comprehensive survey, *IEEE International Conference on Power, Control, Signals and Instrumentation Engineering*, pages 2032–2035, 2017. DOI: 10.1109/icpcsi.2017.8392072 73

[213] T. N. Pappas, New challenges for image processing research, *IEEE TIP*, 20(12):3321, 2011. DOI: 10.1109/tip.2011.2172110 73

[214] I. Amro, J. Mateos, M. Vega, R. Molina, and A. K. Katsaggelos, A survey of classical methods and new trends in pansharpening of multispectral images, *EURASIP Journal on Advances in Signal Processing*, 2011(1):1–22, 2011. DOI: 10.1186/1687-6180-2011-79 73

[215] S. Chikkerur, V. Sundaram, Vijay, M. Reisslein, and L. J. Karam, Objective video quality assessment methods: A classification, review, and performance comparison, *IEEE Transactions on Broadcasting*, 57(2):165–182, 2011. DOI: 10.1109/tbc.2011.2104671 73

[216] P. Marziliano, F. Dufaux, S. Winkler, and T. Ebrahimi, A no-reference perceptual blur metric, *IEEE ICIP*, 3:57–60, 2002. DOI: 10.1109/icip.2002.1038902 74

[217] A. N. Skodras, C. A. Christopoulos, and T. Ebrahimi, JPEG2000: The upcoming still image compression standard, *Pattern Recognition Letters*, 22(12):1337–1345, 2001. DOI: 10.1016/S0167-8655(01)00079-4 74

[218] S. Lee, M. S. Pattichis, and A. C. Bovik, Foveated video quality assessment, *IEEE Transactions on Multimedia*, 4(1):129–132, 2002. DOI: 10.1109/6046.985561 74

[219] J. Rego, K. Kulkarni, and S. Jayasuriya, Robust lensless image reconstruction via PSF estimation, *Winter Conference on Applications of Computer Vision (WACV)*, 2021. 74

[220] M. Gupta, A. Jauhari, K. Kulkarni, S. Jayasuriya, A. Molnar, and P. Turaga, Compressive light field reconstructions using deep learning, *CVPR Workshop on Computational Cameras and Displays (CCD)*, 2017. 74

[221] M. Hirsch, S. Sivaramakrishnan, S. Jayasuriya, A. T. Wang, A. Molnar, R. Raskar, and G. Wetzstein, A switchable light field camera architecture using angle sensitive pixels and dictionary-based sparse coding, *IEEE ICCP*, 2014. 74

[222] O. Iqbal, S. Siddiqui, J. Martin, S. Katoch, A. Spanias, D. Bliss, and S. Jayasuriya, Design and FPGA implementation of an adaptive video subsampling algorithm for energy-efficient single object tracking, *IEEE ICIP 2020*, Oct. 2020. 74

[223] A. Gevins, J. Le, N. K. Martin, P. Brickett, J. Desmond, and B. Reutter. High resolution EEG: 124-channel recording, spatial deblurring and MRI integration methods. *Electroencephalography and Clinical Neurophysiology*, 90(5):337–358, 1994. 74

[224] J. Dutta, G. El Fakhri, X. Zhu, and Q. Li, PET point spread function modeling and image deblurring using a PET/MRI joint entropy prior. In *2015 IEEE 12th International Symposium on Biomedical Imaging (ISBI)*, pages 1423–1426). April, 2015. DOI: 10.1109/ISBI.2015.7164143 74

[225] D. C. Noll, J. M. Pauly, C. H. Meyer, D. G. Nishimura, and A. Macovskj, Deblurring for non-2D Fourier transform magnetic resonance imaging, *Magnetic Resonance in Medicine*, 25(2):319–333, 1992. 74

[226] D. H. Frakes, J. W. Monaco, M. J. T. Smith, and A. P. Yoganathan. Data reconstruction using directional interpolation techniques, U.S. Patent 7,831,088, issued November 9, 2010. 74

[227] Y. Bai, G. Cheung, X. Liu, and W. Gao, Graph-based blind image deblurring from a single photograph, *IEEE TIP*, 28(3):1404–1418, 2018. DOI: 10.1109/tip.2018.2874290 74

[228] Y. Bai, G. Cheung, X. Liu, and W. Gao, Blind image deblurring via reweighted graph total variation, *IEEE ICASSP*, pages 1822–1826, 2018. DOI: 10.1109/icassp.2018.8462255 74

[229] D. I. Shuman, S. K. Narang, P. Frossard, A. Ortega, and P. Vanderghenyst, The emerging filed of signal processing on graphs, *IEEE Signal Processing Magazine*, 30(3):83–98, 2013. DOI: 10.1109/msp.2012.2235192 74

[230] S. Chen, A. Sandryhaila, J. M. F. Moura, and J. Kovacevic, Signal denoising on graphs via graph filtering, *IEEE Global Conference on Signal and Information Processing*, pages 872–876, 2014. DOI: 10.1109/globalsip.2014.7032244 74

[231] M. Zhou, J. Scott, B. Chaudhury, L. Hall, D. Goldgof, K. W. Yeom, M. Iv, Y. Ou, J. Kalpathy-Cramer, S. Napel, et al., Radiomics in brain tumor: Image assessment, quantitative feature descriptors, and machine-learning approaches, *American Journal of Neuroradiology*, 39(2):208–216, 2018. DOI: 10.3174/ajnr.a5391 74

[232] M. Narwaria and W. Lin, SVD-based quality metric for image and video using machine learning, *IEEE Transactions on Systems, Man, and Cybernetics, Part B*, 42(2):347–364, 2011. DOI: 10.1109/tsmcb.2011.2163391 74

[233] S. Wang, C. Deng, B. Zhao, G-B. Huang, and B. Wang, Gradient-based no-reference image blur assessment using extreme learning machine, *Elsevier Neurocomputing*, 174:310–321, 2016. DOI: 10.1016/j.neucom.2014.12.117 74

[234] M. Narwaria and W. Lin, Objective image quality assessment based on support vector regression, *IEEE Transactions on Neural Networks*, 21(3):515–519, 2010. DOI: 10.1109/tnn.2010.2040192 74

Author's Biography

JUAN ANDRADE

Juan Andrade was born in Cuenca, Ecuador on October 25, 1972. He received his B.Sc. degree in electrical engineering from the University of Cuenca. In 2000, he received a Master's degree in mobile communications from Polytechnic University of Catalonia Barcelona-Spain, and his M.Sc. and Ph.D. from Arizona State University in 2005 and 2019, respectively. He is a professor at the University of Cuenca where he is the director of the school of telecommunication engineering. His current research interests include signal and image processing, computer vision, and image understanding.

Printed in the United States
by Baker & Taylor Publisher Services